Game Theory

by Edward C. Rosenthal, Ph.D.

A member of Penguin Group (USA) Inc.

To Bryony, Jordan, and Chloe.

ALPHA BOOKS

Published by the Penguin Group

Penguin Group (USA) Inc., 375 Hudson Street, New York, New York 10014, USA

Penguin Group (Canada), 90 Eglinton Avenue East, Suite 700, Toronto, Ontario M4P 2Y3, Canada (a division of Pearson Penguin Canada Inc.)

Penguin Books Ltd., 80 Strand, London WC2R 0RL, England

Penguin Ireland, 25 St. Stephen's Green, Dublin 2, Ireland (a division of Penguin Books Ltd.)

Penguin Group (Australia), 250 Camberwell Road, Camberwell, Victoria 3124, Australia (a division of Pearson Australia Group Pty. Ltd.)

Penguin Books India Pvt. Ltd., 11 Community Centre, Panchsheel Park, New Delhi—110 017, India

Penguin Group (NZ), 67 Apollo Drive, Rosedale, North Shore, Auckland 1311, New Zealand (a division of Pearson New Zealand Ltd.)

Penguin Books (South Africa) (Pty.) Ltd., 24 Sturdee Avenue, Rosebank, Johannesburg 2196, South Africa

Penguin Books Ltd., Registered Offices: 80 Strand, London WC2R 0RL, England

Copyright © 2011 by Edward C. Rosenthal

All rights reserved. No part of this book shall be reproduced, stored in a retrieval system, or transmitted by any means, electronic, mechanical, photocopying, recording, or otherwise, without written permission from the publisher. No patent liability is assumed with respect to the use of the information contained herein. Although every precaution has been taken in the preparation of this book, the publisher and author assume no responsibility for errors or omissions. Neither is any liability assumed for damages resulting from the use of information contained herein. For information, address Alpha Books, 800 East 96th Street, Indianapolis, IN 46240.

THE COMPLETE IDIOT'S GUIDE TO and Design are registered trademarks of Penguin Group (USA) Inc.

International Standard Book Number: 978-1-61564-055-3
Library of Congress Catalog Card Number: 2010912362

13   8 7 6 5

Interpretation of the printing code: The rightmost number of the first series of numbers is the year of the book's printing; the rightmost number of the second series of numbers is the number of the book's printing. For example, a printing code of 11-1 shows that the first printing occurred in 2011.

Printed in the United States of America

Note: This publication contains the opinions and ideas of its author. It is intended to provide helpful and informative material on the subject matter covered. It is sold with the understanding that the author and publisher are not engaged in rendering professional services in the book. If the reader requires personal assistance or advice, a competent professional should be consulted.

The author and publisher specifically disclaim any responsibility for any liability, loss, or risk, personal or otherwise, which is incurred as a consequence, directly or indirectly, of the use and application of any of the contents of this book.

Most Alpha books are available at special quantity discounts for bulk purchases for sales promotions, premiums, fund-raising, or educational use. Special books, or book excerpts, can also be created to fit specific needs.

For details, write: Special Markets, Alpha Books, 375 Hudson Street, New York, NY 10014.

Publisher: *Marie Butler-Knight*
Associate Publisher: *Mike Sanders*
Senior Managing Editor: *Billy Fields*
Senior Acquisitions Editor: *Paul Dinas*
Development Editor: *Jennifer Moore*
Senior Production Editor: *Janette Lynn*

Copy Editor: *Cate Schwenk*
Cover Designer: *Rebecca Batchelor*
Book Designers: *William Thomas, Rebecca Batchelor*
Indexer: *Heather McNeill*
Layout: *Ayanna Lacey*
Proofreader: *Laura Caddell*

Contents

Part 1:   The Basics 1

1   What Exactly Is Game Theory? 3
   The First Moves 3
   Game Basics 4
      Real Life vs. Mathematical Theories 5
      A Real-World Application 5
   Zero Sum Games 6
      Two Finger Morra 6
      Constant Sum Games 7
   Nonzero Sum Games 8
   Extensive Form Games 10
   Games with Perfect Information 11
   Games with Incomplete Information 12
   Cooperative Games 13
   Solving Games 13
   Consider the Other Player's Position 15

2   Playing the Percentages 17
   The Not-So-Average World of Averages 17
      Simple Averages 17
      Weighted Averages 18
      Probabilities and Expected Values 19
      Maximin and Minimax 21
      Deciding Which Decision Criterion to Apply 22
   Decision Theory 23
      Strategic Choices 23
      The Math of Decisions 24
      In the Real World 26
   Decision Trees 26
      Creating the Decision Tree 26
      Play It Safe or Roll the Dice? 27
   Statistics, Data, and Trustworthy Conclusions 29
      Is That Swab Reliable? 29
      Reliable Tests and Unreliable Results 31

3   Zero Sum Games 35
   Set Up and Solution 35
      Mixed Strategies 36
      Pure Strategies and Domination 36
      The Equilibrium as a Solution 38

	War Games *39*
	Constant Sum vs. Zero Sum *41*
	Solving Zero Sum Games 42
	Mixed Strategy Equilibria *42*
	Solve the Problem for NBC *43*
	Do the Math *43*
	The Analysis *44*
	Solve the Problem for CBS *45*
	Do the Math *45*
	Saddle Up *48*
	Supersized Strategy Sets *49*
	Misconceptions About the Zero Sum Concept 50
4	**Nonzero Sum Games 53**
	Nonzero and Its Rich Flavor 54
	The Coordination Game 54
	Nash Equilibria 55
	Pure Strategy Nash Equilibria *55*
	Mixed Strategy Nash Equilibria *56*
	Equilibrium Payoffs *57*
	Tweaking Payoffs and Shifting Strategy *59*
	Narrowing Down the Solution Concept 60
	The Centipede Game *61*
	Subgame Perfection *62*
5	**More Nonzero Sum Games 65**
	Conflict and Cooperation 65
	The Prisoner's Dilemma 66
	Joint Ventures *69*
	Advertising Wars *70*
	Resource Management *71*
	The Game of Chicken 72
	Stag Hunt 75
	Transform Lose-Lose Outcomes to Win-Win 77
	Strong and Weak Prisoner's Dilemmas *78*
	Hostages and Enforcement *79*

Part 2: How Information Affects Games 81

6 Ignorance Is Not Bliss 83
Perfect and Imperfect Information 83
The Fog of Incomplete Information 85
When Information Is Asymmetric 86
    Fool's Gold *87*
    Beyond Fool's Gold *91*
Breaking Through the Fog 91

7 Signaling and Imperfect Info 97
The Costs of Signaling 97
Separating the Men from the Boys 99
    Do the Math *100*
    The Analysis *100*
Pooling Everyone Together 101
    Do the Math *102*
    The Analysis *102*
Signaling Quality 102
    Selling Lemons *102*
    Market Failure and Insurance Nightmares *105*
Status Symbols as Strategies 106
    Expensive Signals *107*
    The Handicap Principle *107*

8 Nuanced Messages 111
Reading Between the Lines 111
    Signaling to Investors *111*
    Developing a Reputation *112*
    Costly Commitment *113*
Threat Strategies 115
    Brinkmanship *115*
    The Utility of Erratic Behavior *117*
Veiled Strategies 118
    Bearing Outsized Gifts *118*
    Losing Your Cool *119*

Part 3: Getting Ahead by Working Together 123

9 An Overview of Cooperation 125
Bargaining with Two Players 125
Fair Division Problems 127
Cooperative Games 129
Voting and Game Theory 131

10 Bargaining Games 135
Nash's Standard Model 135
Some Reasonable Assumptions *138*
The Nash Solution *140*
Application to Bankruptcy Problems *140*
The Kalai-Smorodinsky Solution 143

11 Fair Division Games 147
I Cut, You Choose, and Related Methods 147
The Steinhaus-Kuhn Lone Divider Method *148*
Other Divisible Procedures *150*
Proportionality and Envy-Free Allocation 151
The Knaster-Steinhaus Procedure *151*
The Adjusted Winner Procedure *155*
Proportional Allocation *157*

12 Cooperative Games 161
Pay Your Fair Share 161
We're All on the Same Team 162
The Shapley Value 164
Do the Math *166*
The Analysis *167*
Other Cost and Profit Examples 168
An Application to Power Generation *168*
Sharing the Profit *170*
Trying to Satisfy Everyone 171

13 Group Decision Games 175
Voting as a Group Decision 175
Veto Power 176
Disappointments and Surprises 178
Pairwise Comparisons *178*
Individual Rankings *179*

　　　　　Problems with Proportionality 181
　　　　　　　Proportionality and Power 182
　　　　　　　Do the Math 184
　　　　　　　Other Power Indices 185
　　　　　The U.S. Electoral System 186
　　　　　Approval Voting 187
　　　　　The Search for a Perfect Method 188

Part 4: Individual Values vs. the Group 191

14 Individual Gain vs. Group Benefit 193
　　　　　The Tragedy of the Commons 193
　　　　　　　The Commons Game with a Low Threshold 194
　　　　　　　The Commons Game with a High Threshold 197
　　　　　　　Ways Out of the Dilemma 198
　　　　　The Volunteer's Dilemma 200
　　　　　The Free Rider Problem 201
　　　　　Some Experimental Results 203

15 Auctions and Eliciting Values 205
　　　　　Types of Auctions 205
　　　　　　　First-Price Auctions 206
　　　　　　　Second-Price Auctions 209
　　　　　　　Common-Value Auctions 210
　　　　　　　Other Auction Types 210
　　　　　Vickrey's Insight 211
　　　　　Auctions in Practice 212
　　　　　　　Misuse of the Vickrey Auction 212
　　　　　　　Awarding TV Rights 213
　　　　　　　Auction Successes 214

16 Designing Games for Group Benefit 219
　　　　　Overcoming Individual Temptations 219
　　　　　　　Adverse Selection 220
　　　　　　　Moral Hazard 221
　　　　　Design for Truthful Revelation 222
　　　　　　　Cost-Sharing Schemes 222
　　　　　　　Vickrey-Clarke-Groves Mechanisms 226
　　　　　　　Do the Math 228
　　　　　　　The Analysis 229
　　　　　　　Internet Advertising 229

        Limitations of Mechanism Design 231
            Efficiency vs. Budgeting *231*
            Computation and Collusion *233*

Part 5: Behavior in Games 235

17 Biology and Games 237
    Hawks, Doves, and Their Strategies 237
    Evolutionarily Stable Strategies 240
    Commitment and Other Virtues 242
    Altruism and Reciprocity 243
    The Excitable Brain 245
        Emotion Trumps Rationality *245*
        fMRI Studies, Choices, and Neuro-What? *246*
    What Future Research Could Mean 249

18 Aligning Theory with Behavior 251
    Utility Theory 251
        The Standard Economic Perspective *255*
        Some Confounding Paradoxes *257*
    Prospect Theory 258
        Gains vs. Losses *259*
        Relative Value *262*
        Risky Business *263*

19 Behavioral Decision Theory 265
    Frames of Reference and Mental Accounting 265
    Anchoring and Relativity 267
    Intertemporal Choice 270
    The Endowment Effect 273

20 Strategic Behavior in the Lab 275
    Dictators and Trust in the Laboratory 275
        The Dictator Game *276*
        Trust in Stages *277*
    Behavior in the Ultimatum Game 281
        Baseline Results *282*
        More Penetrating Findings *283*
    Coordination Experiments 285

21	**More Quirky Behavior** **289**	
	Dominant Strategies in Practice 289	
	Mixed Strategies in Practice 291	
	Beauty Contests and Other Minds 292	
	Prisoner's Dilemma Experiments 294	
	Early Findings *295*	
	Later Discoveries *296*	
	Our Limited Abilities 298	
22	**Repeated Games and Tendencies** **301**	
	Why Study Repeated Games 301	
	The Chain Store Paradox *302*	
	Repeated PDs *305*	
	Some Experimental Corroboration *306*	
	The Folk Theorem 309	
	How Cooperation Can Evolve 310	
	Axelrod's Experiment *311*	
	Building Future Relationships *315*	

Appendixes

A	Glossary 317	
B	Sources, Resources, and Further Reading 329	
	Index 341	

Introduction

In our increasingly hectic world, we are required, even forced, to make decisions all the time. This can be stressful, and we often worry whether we made the right choice.

That's where a little knowledge of game theory comes in handy. Game theory is all about strategizing. It's about making the best decisions you can—how to go about choosing what to do in the face of uncertain outcomes and scheming adversaries.

How do *you* make decisions? Do you make snap judgments? Do you painstakingly note down all the pros and cons and weigh them up? Do your decisions affect others? (Do you care?) And one more question: do you anticipate the decisions that others are making and how they impact you? That last question—about anticipating others' decisions—is what game theory is really about.

Game theory took off around 1950 as a mathematical theory of conflict. It treats situations where there are multiple decision makers, or "players," each of whom has a variety of different alternatives, or "strategies," to employ.

When each player selects a strategy, there is some joint outcome to the game, for example, his or her monetary payoffs. What game theorists do is to figure out how best to play; in other words, how to go about picking the best strategy.

Who uses game theory? As I point out at times, many of the ideas in this book have been striking enough and important enough to warrant a number of Nobel prizes. Economists, biologists, and political scientists have been using game theory for years. But the bigger news for the rest of us is that game theory has been catching on, and helping businesses and governments strategize in a whole host of ways. As you'll also see, game theorists and neuroscientists have recently been teaming up and are finding out what makes us tick.

Playing games and, especially, winning and losing are central to our culture. Many of us are rabid sports fans who get a primal kick out of seeing our team win. Being called a "loser," on the other hand, is just about the worst insult imaginable. Gambling is increasingly becoming a national pastime. But more telling is that poker has become a "sport" we can watch on TV. It combines the competitive and monetary elements of sports and gambling with a cerebral dimension that many of us also crave.

Game theory has the same fundamentals to it: competition, money, and guile. One enduring lesson, however, that I stress is how cooperation keeps emerging—surprisingly, to everybody's advantage—even in competitive situations.

One thing that sets this book apart from other game theory books is its psychological angle. If you want to get ahead (or just get along) in your personal lives as well as in the business world, you need to understand not just how people ought to make decisions, but how they actually behave. This book provides plenty of insights into both.

How This Book Is Organized

Part 1, The Basics, covers all the essential stuff you need to know to get started. Since the world of human interaction is complex and poses a number of different types of situations where we have to think strategically, game theorists have developed a variety of different game models in response. So first off, I walk you through all of the different models and how they work.

Once you understand how the different games are set up, you need to know the fundamental concepts regarding how best to play. These concepts involve what it means to solve a game, and how to get to those precious solutions.

The rest of Part 1 introduces the basic tools necessary for working through solutions and then applies those tools to the classic types of zero sum and nonzero sum games. These tools and solution concepts give you a vocabulary and a thought process to ease your understanding of everything else that follows.

Part 2, How Information Affects Games, shows you that while too much information is not a problem in the world of game theory, not knowing what the other player is getting out of the game can really handicap your strategizing. You'll learn how to avoid getting badly fooled when things are hidden from you.

I also explore the significance of the signals that people send all the time. What does your college degree signify? Should you be suspicious of that used car? Why do we go out of our way to spend lavishly?

Finally, some signals are more understated than the Ivy League sweatshirt. I review what game theory has to say on signaling your ability and your trustworthiness to your clients, investors, and friends, and how to structure delicate, risky situations.

Part 3, Getting Ahead by Working Together, explores one of the central revelations of game theory: more often than you might suspect, cooperation and competition are inextricably linked. When people join together in an enterprise, what is the synergy that develops and how should the costs and profits be shared? No one wants to pay too much or receive too little.

In these joint decisions, which include bargaining, dividing up spoils, cost sharing, and even voting, it's all about being fair and I show you how game theory provides a way to implement fair solutions in a variety of these cooperative, but still competitive, situations.

Part 4, Individual Values vs. the Group, highlights another critical element in game theory, which is the conflict between what people want for themselves and what is best for the group. After illustrating the clash between individual desire and group success in a variety of settings, I show what game theory can contribute to this enduring problem.

One area of interest here is the study of auctions, which often force people's hand when considering their preferences among different alternatives. But the key to Part 4 is a concept called mechanism design. This is a way to engineer the rules of the game so players will still respond naturally to incentives, and yet to steer those incentives in the direction of the common good. Of course, this sounds too good to be true, and I discuss some unavoidable limitations of this approach.

Part 5, Behavior in Games, covers some hugely important developments that hardly appear, if they appear at all, in other game theory books. Game theory started out as a mathematical science purporting to show how rational decision makers ought to act in certain, well-defined situations. A small detail—that those decision makers are fallible human beings—has been making game theorists uncomfortable for the past 60 years. Not surprisingly, many authors just sweep this issue under the rug.

About 30 years ago, however, some brave souls started to speak up and show everybody that people are not the cold and precise calculating machines that the pure underlying theory assumes.

We need the theory, since it provides the "rational" yardstick for us to measure up to, but we also need to know what human beings really comprehend in risky situations and how they actually strategize. As you'll find out, there is a method to our madness, and quite often there is a benefit, too.

Two appendixes offer additional resources. In Appendix A you'll find an extensive glossary that provides a quick way to keep up with the many specialized terms that are used in game theory. And for those of you who want to dig deeper, Appendix B not only points you to the original sources of much of the material in this book but also recommends some interesting websites and options for further reading.

Extras

I've strategically placed four different kinds of sidebars throughout the book to sharpen your understanding of the many different topics under the umbrella of game theory. The sidebars serve different functions, as described here:

DEFINITION

Key words and phrases necessary to understand the basic game-theoretic concepts.

POISONED PAWN

Advisories that head off misunderstandings and faulty strategizing.

BET ON IT

Clarifications that take your understanding to the next level.

NO BLUFFING

Quirky tidbits that take you in unexpected directions.

If game theory is about anything, it's about anticipating how others will act. Thinking about the others' strategies is to game theory as keeping your eye on the ball is to playing sports. Right now I'm feeling a nervous excitement as I anticipate your getting started devouring this fascinating subject. I've thoroughly enjoyed writing this book and hope you have the same fun going through it.

Acknowledgments

I learned game theory from Ehud Kalai and Barry O'Neill at Northwestern University, and want to thank them for helping shape my interests. Barry in particular inspired me with his unwavering focus on what's fair and what isn't.

My colleague at Temple University, Fred Murphy, has tirelessly reminded me from day one that games lurk everywhere. "What's the game," is something he's always thinking and I'm grateful to have that reminder. I also want to acknowledge the help of another Temple colleague, Mike Powers.

I want to thank my agent, Verna Dreisbach, and my acquisitions editor at Alpha Books, Paul Dinas. In the early going, Paul got me focused on how I needed to craft this book and he's been there for me throughout the process. Thanks also to Jennifer Moore for the great job she did editing this book, and to Janette Lynn and Cate Schwenk for their work later on in the process.

I must think of a way to thank my wonderful wife, Bryony Kay, for her critical eye, unending support, and valuable feedback. Thanks also to our great kids, Jordan and Chloe, for understanding that I've been pretty busy these past months and for putting up with my extended absences in the back room.

One more colleague at Temple University needs a special acknowledgment. Dimitrios Diamantaras is a game theorist in our economics department who has an awesome command of the subject. I am very fortunate that Dimitrios cheerfully agreed to come on board as the technical editor. His careful reading and subtle insights have improved this book immensely.

Trademarks

All terms mentioned in this book that are known to be or are suspected of being trademarks or service marks have been appropriately capitalized. Alpha Books and Penguin Group (USA) Inc. cannot attest to the accuracy of this information. Use of a term in this book should not be regarded as affecting the validity of any trademark or service mark.

Part 1

The Basics

Now it's time to reveal what game theory is all about. I first introduce the different types of games and what purposes they serve. For instance, zero sum games are the most basic games, in which one player gains what the other loses. Most situations in the world, though, are too complex to be captured by zero sum games. When there's synergy, for example, both players can profit, and these situations call for nonzero sum games.

There are many other game models, too. Games vary according to the amount of information the players have, and I introduce the various ways in which the available information changes the picture.

Once you have these basic game structures under your belt, I show you how to solve the games using some basic math. The solutions are called *equilibria* because the players arrive at a point where their outcomes can't be improved by any different choice of strategy.

What Exactly Is Game Theory?

Chapter 1

In This Chapter

- What game theory is (and isn't)
- A few basic games
- How to begin solving games
- The keys to game theory

Game theory is a mathematical approach to real-life situations that involves two or more decision makers. Each decision maker—you, for example—has a number of different actions available, and the ultimate outcome depends not only on your action but also on what the others do. As a consequence, the decision makers are forced to think strategically, and it is this feature that gives game theory its intrigue. In this chapter I introduce you to the basic game theory concepts you'll need to know.

The First Moves

Game theory has its origins in the 1920s when John von Neumann (1903–1957), a brilliant Hungarian mathematician, wanted to figure out a scientific approach to bluffing in poker.

Von Neumann realized that playing poker requires a strategic element that games like chess lack. In most ordinary games the players can see all the pieces on the board. But in poker you don't know what cards the other players have, and this is what gives players the chance to misrepresent their strengths and weaknesses. Von Neumann realized that a science of how to best act in these situations would apply to a huge swath of decision-making in business and society.

> **BET ON IT**
>
> Von Neumann was a genius among geniuses, having made pioneering contributions to a wide range of fields in math and physics. He was also a key member of the Manhattan Project, the group of scientists responsible for developing the first atomic bomb. But what made von Neumann legendary was his more wacky side: his storytelling, his partying, his circus feats of the mind. Who else could help develop the Bomb, but also play practical jokes on Einstein?
>
> In 1933, von Neumann was one of the original five professors hired at the Institute for Advanced Study at Princeton. Over the next quarter century he put his unequalled talent and inexhaustible energy toward all manner of scientific projects. But looking back now, the single most outstanding contribution he made was to pioneer game theory.

Other mathematicians studied games before von Neumann, but no one else developed their work into an applied science. Somehow finding time outside his huge contribution to the allied war effort, von Neumann wrote the first book on game theory with economist Oskar Morgenstern in 1944.

Around 1950 a whole flurry of research in game theory surfaced at places like Princeton and the RAND corporation (a well-known think tank). In the 1970s game theory began to spread from its mathematical roots and become a useful and accepted methodology in areas like economics, political science, and biology.

In the 1990s the Nobel committee began to award Nobel prizes to game theorists for their contributions to economics. Today people in the mainstream are starting to realize that game theory offers insights into other, more practical, pursuits like business strategy.

Game Basics

A game is a situation in which there are multiple decision makers. Each decision maker is called a player, and each player has a certain set of actions, called strategies, available to him or her. When each player settles on a particular strategy, the result is called the outcome. These outcomes are measured numerically and are often referred to as payoffs.

Game theory refers to an ever-growing set of *mathematical models* and solution procedures that are intended to inform us about what the players should do when they need to act in a particular setting. Game theory is normative, or prescriptive: it tells us the "best" action to take, based on a reasonable set of principles. I've put quotes around the word "best" because sometimes the notion of best is problematic. You'll soon understand why.

> **DEFINITION**
>
> A **mathematical model** is the use of variables, equations, probabilities, and so on, in order to represent a real-world situation.

Real Life vs. Mathematical Theories

Although game theory can guide our decisions, keep in mind that in real life, people in game situations often do not behave according to the mathematical theory, which explains why it's difficult to use game theory as a predictive tool regarding human behavior. This additional twist explains why we need to balance the theory of what we *should* do with an understanding of how people *actually* behave. I introduce you to the descriptive side of game theory in Part 5.

Game theorists have developed a number of different types of games as a way to deal with the complexity of the world that we live in. Some situations are purely competitive; some are literally war; some require coordination between different parties. In some games the players move in a certain sequence; in some games the players are all on the same team; and in other games, you don't know anything about the players you're up against.

A Real-World Application

With a basic understanding of games under your belt, let's ease into our very first mathematical model.

Suppose you and your friend go out to dinner on Friday nights. To make it interesting, each of you secretly writes down a choice of restaurant (say, Thai or Italian). To decide where you'll be going in the fairest, most objective, way, you agree to flip a coin. The mathematical model here is simply the use of a coin flip to randomly pick a winner while giving each player an equal (50 percent) chance at victory.

Of course, it's possible for random play to be unfair. For example, if your friend rolls two dice and you get your pick of restaurant only on double sixes, the situation is clearly tilted against you. When you casually roll dice or flip coins, you fully understand that there is a winner and a loser.

Zero Sum Games

Zero sum games are entirely competitive. One player gains what the other loses. But there is a subtlety to these situations that we need to uncover.

> **DEFINITION**
>
> A **zero sum game** is a game with two players in which every possible pair of numerical outcomes adds up to zero. In zero sum games, what one player gains, the other loses.

Two Finger Morra

Many of us played the "evens" or "odds" game of Two Finger Morra as children. In the game, one player calls either "evens" or "odds." Both players then stick out one or two fingers at the same time. The player who calls "odds" wins if one of the players has one finger showing while the other has two fingers out (an "odd" result). If both players' fingers match, the player who had called "evens" wins.

Two Finger Morra is an example of a zero sum game. In every possible scenario, one player wins while the other loses. The usual way to depict these games is to draw a table, often referred to as a matrix in game theory, which shows all of the strategies available as well as each outcome. This is called representing a game in *strategic form* or normal form.

> **DEFINITION**
>
> A game is in **strategic form** when a table is used to enumerate the strategies for the two players as rows and columns, and the entries in the table indicate the outcomes.

The potential outcomes in the game are represented in the table by a pair of numbers in each cell: each pair consists of a 1 to represent victory and a -1 to represent defeat. Notice how adding together 1 and -1 equals zero, as in "zero sum."

Suppose you had called "evens" prior to playing Two Finger Morra. Your strategies are to throw out either one finger or two fingers. These choices correspond to the two rows of the following table. Had you thrown out one finger, the resulting outcomes are listed in the first, or top, row of the table; had you thrown out two fingers, the corresponding outcomes are shown across the second, or bottom, row.

Similarly, your opponent has two strategies. Their choice of throwing out one finger yields the outcomes in the first, or leftmost column, and their throwing out two fingers yields the outcomes in the next column. In each cell in the table the row player's payoff (in this case, yours) is followed by the column player's payoff.

		The Opponent's Moves	
		One	Two
Your Moves	One	1, -1	-1, 1
	Two	-1, 1	1, -1

Two Finger Morra represented in strategic form.

NO BLUFFING

The expression "zero sum game," has become synonymous with trade-offs in general and with finite resources in particular. It is a favorite media expression used to describe important policy topics like trading off one tax break against some other tax increase.

Constant Sum Games

Zero sum games can get a lot more complex. Let's consider a zero sum game with more at stake than choosing sides on the playground. Suppose you are in charge of evening programming for a major television network (let's call your company Network A), and you are trying to figure out what sort of prime-time program to run in the 9 P.M. Saturday prime-time slot. You have a lot of competitors—three other major networks as well as dozens of cable channels—but you are most concerned about a particular competing network, Network B.

Suppose that both Network A and Network B have narrowed their choices down to running either a sitcom or a sports program. Suppose further that historical information combined with up-to-date estimates predict the following market share figures as shown in the following table. Each entry shows A's market share followed by B's market share.

		Network B	
		B's Sitcom	B's Sports
Network A	A's Sitcom	56%, 44%	47%, 53%
	A's Sports	46%, 54%	48%, 52%

Network market share.

Notice that Network A's sitcom beats Network B's sitcom, but in the other three scenarios, Network B comes out on top. Notice that this game isn't as easy to resolve as Two Finger Morra, and it doesn't seem as fair, either. If you're paying close attention, you might also notice that the market share percentages do not add up to zero. In each location in the table the percentages add up to 100 percent, not zero. So how can this be a zero sum game?

In fact, the network programming game is a constant sum game that represents an unchanging pie. In such games, each set of outcomes adds up to the same number, indicating a constant total—but that number isn't necessarily zero. Constant sum games are intertwined with zero sum games and, as such, are not a separate category of games. I tease out these distinctions in Chapter 3.

Nonzero Sum Games

Nonzero sum games are set up just like zero sum games, but the numerical outcomes do not always add up to zero or to a given constant the way they do for zero sum or constant sum games. In nonzero sum games the outcome pairs add up to different amounts, which means that one player's gain is not necessarily the other's loss.

One type of nonzero sum game is the Coordination game. In coordination games, each player is rewarded when their strategies match but they both suffer when their strategies don't match. To see how the Coordination game plays out, consider the case of two electronics giants, Sony and Toshiba, which have each developed a new technology. Their technologies are extremely similar but require different equipment.

Sony and Toshiba would like to end up using compatible equipment. They believe that compatibility will not only generate synergy for their products but will discourage competitors like Samsung from marketing a similar technology. However, for either company to adopt the other's format would require a significant investment.

To come up with a matrix for this game, you'll need to assign reasonable values for each player for every possible outcome. Sometimes these values denote monetary figures, such as costs or profits. Sometimes they reflect other information, such as the market shares or technology adoption we've just considered.

Often, the numerical values simply represent relative preferences, not too different from when you are asked to rate an experience on a scale from 1 to 10. Game theorists call these numerical values *utilities*, to signify how much the players will gain from a particular result.

DEFINITION

Utilities are measures of one's relative gain, pleasure, or satisfaction from a certain outcome.

Let's develop the matrix of outcomes. Sony's technology is Blu-ray, while Toshiba's is HD DVD. Sony prefers the Blu-ray format to HD DVD, but would adopt HD DVD rather than not coordinate at all. Toshiba's preferences are equal in strength, but opposite to Sony's. In other words, Toshiba would prefer to adopt HD DVD over Blu-ray, but would adopt Blu-ray rather than not coordinate at all. The table of Sony's and Toshiba's respective utilities, also called a payoff matrix, is included here. Higher numbers indicate stronger preferences, with zero indicating the least desirable outcome.

		Toshiba's Choices	
		Blu-ray	HD DVD
Sony's Choices	Blu-ray	(4, 1)	(0, 0)
	HD DVD	(0, 0)	(1, 4)

A payoff matrix for the Coordination game.

Based on the matrix, if Sony and Toshiba both pick Blu-ray, Sony is pleased (utility of 4), and Toshiba less so (utility of 1); on the other hand, if Sony insists on Blu-ray and Toshiba sticks with HD DVD, they end up failing to coordinate, and are not only irritated with each other but worried that Samsung will enter the picture. This mutual dissatisfaction is indicated by the (0,0) outcome.

Sony can yield to Toshiba, too. If they both pick HD DVD they obtain the reverse of the Blu-ray outcome. But ironically, if they both give in to each other, they end up in the bottom left of the table, remaining incompatible.

The Coordination Game is an example of a nonzero sum game because the different outcome pairs add up to different amounts. When the two firms coordinate, their outcomes add up to 5, while a failure to coordinate yields a sum of 0.

Extensive Form Games

Some situations are more complicated than simply listing outcomes in rows and columns. In certain situations, players have to move in a specific sequence. In other situations some players are at a known position while other players are not sure where they are. It is often useful to represent these complexities by drawing a diagram.

These diagrammatic representations are called *extensive form games.* They begin with a single point, or node. Each node, represented by a circle in the diagram, indicates when a player has a decision to make. Different moves, or actions, are represented by branching lines that fan out (from left to right) from the starting node. Each branch leads to another node, at which point either another decision takes place, or else the game ends with a certain payoff. Let's take a look at an example.

DEFINITION

An **extensive form game** is represented by a branching diagram in which the branches represent alternatives available to the players.

Suppose you are watching a soccer game and a player, K, is about to take a penalty kick. The ball sits on the ground and K is about to run up and belt it. The tense goalkeeper G is focusing on which way the ball will go.

Typically, to maximize the scoring chance, K will kick toward either the right or left corner. Aware of this, G will dive right or left. If G dives the wrong way he has no chance of stopping the ball (of course, K might miss anyway), but if G guesses correctly he has some hope of blocking the kick. This situation is represented in the following extensive form diagram.

The first thing to notice is that the game appears to be sequential. It seems as if K makes her move and then G responds. In a real game, G's dive is pretty much simultaneous with K's kick. Since G does not know which way the ball is going to go, we need a way to show this in the diagram.

To overcome this confusion, the two nodes representing G's move are both shaded gray. This means that they are indistinguishable to G. Game theorists call such a set of nodes an *information set*.

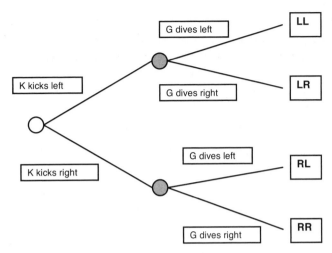

A penalty kick game in extensive form.

At an information set, the player picks a strategy without knowing exactly which node she is on in the game tree. In other words, her information at that point in time is identical at all nodes in the information set.

To fully describe the game, we need to assign values to the outcomes LL, LR, and so on, in the diagram. Evaluating these outcomes would involve factors like the probability that K will miss when kicking to her right, how well G has defended other penalty kicks in the past, and so on.

POISONED PAWN

You might be wondering whether formulating the penalty kick game in extensive form is better than simply making up a table. When games really are sequential, an extensive form diagram is the best way to go, since the branches show the flow of the game from one player's move to the next. There is no easy way to show sequential moves using tables.

Games with Perfect Information

In games like chess (and checkers, Go, and tic-tac-toe), every information set is a single node. In other words, nothing is hidden, so every time one player makes a move, they both know the resulting configuration. Such games are called games of perfect information because both players know exactly where they are at every play.

> **BET ON IT**
>
> Some perfect information games have nothing to do with board games. For instance, in a principal-agent game an employee's work is evaluated by management. The employee, understanding that the bonus he will receive depends on his level of effort, will choose how hard to work. Then management observes the employee's output and awards the appropriate bonus. What game theorists call perfect information is called "transparency" by management.

Games with perfect information range from the fairly simple to the enormously complex, in which the game trees are too big to solve for even the fastest computers. In chess, for example, there might be a couple of dozen possible moves at every turn. Looking only eight moves ahead would generate a game tree with more than a trillion branches to evaluate!

Games with Incomplete Information

The idea behind games with incomplete information is that even if you can narrow down the strategies of the other players, you don't know what their outcomes are. In other words, in games with incomplete information, the players might know their own outcomes but do not know the outcomes for the other players.

Think about a contract negotiation between a labor union and the management of a large company. Even when discussing tangible items like health insurance co-pay, the labor union does not know the true cost of the co-pay plans to the employer. This lack of information makes it difficult for the union to know just how onerous a particular plan is to the employer.

With incomplete information it becomes very difficult to strategize because you don't know your opponent's true nature. In a very real sense, since you don't know some of the outcomes, you don't even know what the game is that you're playing.

With games of incomplete information the solution involves trying to gauge the other players' utilities. In other words, each player attempts to size up the monetary outcomes or, at least, the relative preferences, of the other players in each of the different outcomes. These unknown evaluations are called types (as in, what type of folks we are up against). For example, in the contract negotiation, the labor union would try to guess whether management would truly be unable to accommodate a certain amount of health co-pay. In this case the union is simply sizing up management as being one of two types: either able to afford a certain premium, or not.

Cooperative Games

Cooperative games are used to model situations in which players are better off when they join up with others. Cooperative games typically have more than two players, and even though there is still underlying competition in these situations, the players will nevertheless benefit by forming subsets called coalitions.

Usually in cooperative games, the costs or benefits obtained by the different possible coalitions can be measured in monetary terms. And typically, these monetary costs or benefits end up getting divided by the different players in the coalitions. For this to occur, there must be some way for money to change hands. Games with this feature are called games with side payments. Some cooperative games, however, model situations where there is no good way for the benefits to be divided up and shared. These games are called nontransferable utility games.

To represent these games, you must list all of the possible coalitions that can be formed, together with what each coalition is able to accomplish by way of costs or profits, when cooperation is limited to just that coalition. In other words, each coalition's value is determined solely by what they are able to accomplish on their own, without the participation of the remaining players.

Here's an example: you and your three roommates sign up for a cable television plan that costs $86.17 a month. You could share the cost equally but realize that you don't all watch equal amounts of TV. On top of that, you've all insisted on different premium channels, which helped drive the cost up. What's a fair way to split the monthly charge?

Solving Games

You've just learned a sampling of the major categories of games within game theory and some of the ways those games are structured and organized. So now I hope you're wondering how you can use game theory to actually help find a solution to any particular game.

I mentioned earlier that sometimes the notion of "best" play is elusive. One way to decide the best strategy is to compare the average payoffs of alternative strategies. Another option is to minimize your exposure to negative outcomes. These criteria, however, are not employed in a vacuum; you will always have to include your opponents' possible actions (and their payoffs) in your calculations.

The solutions that we'll come up with are recommendations on how to act in the various game situations. If you were going to play Rock-Paper-Scissors and asked me what strategy you should employ, I would advise you to randomly play rock a third of the time, paper a third of the time, and scissors a third of the time.

> **POISONED PAWN**
>
> Sometimes the recommended play requires you to act randomly according to certain probabilities. You might disdain the idea of carrying out an important decision by flipping a coin, rolling dice, or employing some other kind of randomizing strategy, but try to resist this urge. Oddly enough, using randomness in a calculated way is sometimes the most logical thing to do.

Such advice might frustrate you. You'd rather hear a more concrete instruction like, "play paper." In fact, if you were playing your six-year-old nephew, I might suggest you play paper, since he might be somewhat inclined to play rock. In fact, statistically speaking, people out there in the world do seem to play with a slight bias—meaning that their actual plays are not done 33.3 percent each. If you knew their bias, you could adjust your play accordingly.

As you will discover, the solution methods vary according to the information that is available to the players. The following table illustrates how the information available to the players differs among the games you have learned about so far.

Game	Outcomes
Two Finger Morra	Win/lose
Market Share game	Known market shares
Coordination game	Payoffs known but vary; the size of the pie is no longer fixed
Contract Negotiation	Don't know what a policy is worth to others

Sometimes the "best" thing to do depends on whether you are playing the game just once or whether it's a repeated situation. (Are you dining at the Thai or Italian place just this one time, or do you face the same challenge every Friday?) And sometimes the best strategy when facing a "rational" opponent will fail when the adversary does not behave rationally.

Consider the Other Player's Position

Let me wrap up this chapter with the single most important idea in game theory: *always put yourself in the other player's position.* This is a good guiding principle—and one often overlooked—for anyone interested in game theory.

The trick is not to ask yourself what *you* would do in their position, but to ask what *they* would do. Game theory in its pure form treats all players as rational, but people are often irrational in certain, often predictable, ways. Game theory enables us to discover what cold, calculating rationality is—and isn't.

The Least You Need to Know

- Games have players, strategies, and outcomes.
- Zero sum games are purely competitive.
- Nonzero sum games capture cooperative as well as competitive situations.
- Solutions to games depend on mathematical analysis, including probabilities and randomness.

Playing the Percentages

Chapter 2

In This Chapter

- Calculating expected values
- The best of the worst
- Using decision theory to maximize payoffs
- Determining the reliability of information

In game theory, as in life, many events that directly impact us are completely out of our control. Many such events randomly occur and are not brought about by a willful opponent. Game theorists call these events states of nature and have developed strategies for taking them into account. Learning how to strategize in the face of these random events enables you to plot against a thinking adversary. When dealing with these situations, the first step you need to take is to learn how to deal with averages.

The Not-So-Average World of Averages

The innocent-sounding term "average" is actually a complex concept with multiple definitions and implications. Averages take some time to fully grasp, and you need a solid understanding of them because they are the most important tools in your arsenal.

Simple Averages

Most people treat the term "average" as a shortcut to describing some sort of group characteristic. You probably already know how to compute simple averages: you add up all the figures and divide the sum by the total number of terms. If the eight entrées at your favorite little café cost $18, $24, $15, $20, $19, $25, $16, and $19, the average price is: (18 + 24 + 15 + 20 + 19 + 25 + 16 + 19) divided by 8, which equals $19.50.

But what if you are the café owner and someone asks you what your average revenue is from your eight entrées? As the owner, you know that the different dishes are not equally popular. Suppose no one, in fact, ever orders the $25 plate, but the $16 entrée is wildly popular. Given these facts, you know that the average entrée fetches less than $18.

Weighted Averages

The preceding point illustrates that averages depend on frequency, or likelihood, of the given figures, and not just the figures alone. The first way we calculated the average cost is called performing a *simple average*. When you factor in the frequency of the figures, you calculate the *weighted average* instead.

> **DEFINITION**
>
> A **simple average** is an average that counts each figure in the set exactly once.
> A **weighted average** takes into account how frequent, or likely, the figures are.

Suppose you, as the restaurant owner, were analyzing exactly where your revenue was coming from. Let's stick with the eight dishes and their corresponding prices in the preceding section. The data in the following table shows the previous 100 dishes ordered.

Dish	1	2	3	4	5	6	7	8
Price ($)	18	24	15	20	19	25	16	19
Frequency	13	7	11	8	11	0	42	8

Frequency distribution of the previous 100 dishes ordered at your restaurant.

> **BET ON IT**
>
> From this point on I use the symbols "*" and "/" to denote multiplication and division, respectively.

Computing the weighted average revenue is a relatively simple matter: for each entrée you multiply its price by the number of times the dish has been ordered. You then add these terms up over all of the entrées and divide by the total number sold. So the average price the entrées have been bringing in (again, this is the weighted average) is: (18*13 + 24*7 + 15*11 + 20*8 + 19*11 + 25*0 + 16*42 + 19*8)/100, which is equal to $17.60. Notice that this weighted average is a lot lower than the simple average of $19.50.

In this example, the simple average is a useful tool for a diner to get a feel for how expensive the dishes are, but the weighted average is more relevant for the owner when determining average revenue per entrée.

Probabilities and Expected Values

Now it's time for an important alternative framing of the term "average." The calculation you just performed to get the weighted average is simply a shortcut for adding up the prices of all 100 dishes sold and dividing that sum by 100. But now I am going to ask you to make a conceptual leap by taking those frequencies and using them to obtain likelihoods, or probabilities, which each particular dish *will be* ordered.

In the restaurant example the conversion from frequencies to probabilities is easy, since the total number of dishes served is equal to 100. You convert the frequencies into relative frequencies by dividing each one by 100, using the result to represent probabilities. This conversion is as follows:

Dish	1	2	3	4	5	6	7	8
Frequency	13	7	11	8	11	0	42	8
Probability	0.13	0.07	0.11	0.08	0.11	0.00	0.42	0.08

In the table the probabilities merely look like a different, but equivalent, way to represent the frequencies. But the difference in interpretation is crucial. The frequencies represent the *actual* outcomes over the past 100 diners, but the probabilities represent *likelihood* of future outcomes!

This difference is not just semantic. In game theory, it is the difference between what someone has done *up until now*, and what someone might do *from now on*. Since we don't know what is going to happen in the future, and game theory is all about preparing for future encounters, probability is more important than frequency.

The next point requires some mental wrestling. You have to distinguish averages from likely outcomes. Here's an example to get your feet wet: suppose you have some money to invest and you are thinking about putting it into a mutual fund called GROWTH for one year. When you investigate GROWTH's annual performance over the past 10 years, this is what you find:

    6 out of 10 years: up 28% per year

    1 out of 10 years: down 1% per year

    3 out of 10 years: down 12% per year.

If you are willing to base your decision on the activity over the past 10 years, you would convert the frequencies of the three different outcomes to probabilities. Let's use the notation P{event} to represent the probability of a certain event. So:

    P{GROWTH goes up 28% in one year} = $6/10$ = 0.6

    P{GROWTH goes down 1% in one year} = $1/10$ = 0.1

    P{GROWTH goes down 12% in one year} = $3/10$ = 0.3.

Since the frequencies have already been converted to probabilities, to compute the (weighted) average, you just multiply each outcome by its associated probability and add all the terms up. This weighted average is called an *expected value*.

DEFINITION

When probabilities of the outcomes are available, the weighted average, or **expected value,** is found by multiplying each outcome by its associated probability and adding up all the terms.

GROWTH's average annual performance is calculated as follows:

    (28% * 0.6) + (-1% * 0.1) + (-12% * 0.3) = 13.1%.

In other words, the average annual return with GROWTH is 13.1 percent. That probably sounds pretty good to you.

Before calling your broker, let's probe this result a bit. You are wondering whether you should invest your money in GROWTH. Thirteen percent is tempting. Even more tempting: most of the time, you get 28 percent! These are attractive numbers.

But hold on there … it's time to look at the downside. There is a chance you will lose 1 percent. That's not a catastrophe, and it's not particularly likely (0.1 probability) either. But 30 percent of the time you lose 12 percent of your money. That's a big hit, and a somewhat likely one, too!

By now you're probably not so sure about GROWTH. Probing further, you also realize that the 13.1 percent average is never going to be the actual outcome, or anywhere even close. Sixty percent of the time you'll do great, raking in a 28 percent return, but the other 40 percent of the time you will lose money, often a lot of money. But you never actually see a 13.1 percent return. GROWTH leads to either rags or riches, and for many people the risk here might outweigh the return.

How can you reduce your risk? By using one of the two principles outlined in the following section.

Maximin and Minimax

If investing with GROWTH is giving you cold feet, it's because GROWTH's worst-case scenario is pretty bad. What if you found a different fund with no real downside? An investment guaranteed not to lose money sounds attractive.

Suppose you find a bond fund, called BONDD, which has had the following performance over the past 10 years:

    8 out of 10 years: up 4% per year

    2 out of 10 years: up 2.5% per year.

First, to compare BONDD to the previous fund, GROWTH, you compute the average annual return. This expected value is:

    (4% * 0.8) + (2.5% * 0.2) = 3.7%.

Fund BONDD's 3.7 percent return on investment doesn't take your breath away. But while there's no adrenaline, you may have noticed that, unlike GROWTH, BONDD never loses money.

Ask yourself the following question: with BONDD, what is the worst that could happen to you? You look at the possible outcomes (not the average) and see that the lowest annual return is a gain of 2.5 percent.

With GROWTH, you might lose 12 percent. If you want to preserve your wealth, you will reduce your downside by selecting fund BONDD.

You just applied a different *decision criterion* to your investment problem. Instead of going for the fund with the highest average payoff, you selected the fund with the best worst-case outcome. This conservative technique is called maximin: taking the best of the worst-case investment outcomes is identical to taking the *max*imum (best) of the respective *min*imum (worst-case) payoffs.

DEFINITION

A **decision criterion** is a rule you use to make a decision. So far you have learned to use both expected values (weighted averages) and maximin (best of the worst) decision criteria.

To recap, here are the steps involved in performing a maximin analysis:

1. Identify all possible choices that you can make.
2. For each choice, find out what all the possible outcomes are and locate the worst possible choice (the one with the minimum gain).
3. From this set of worst-case scenarios, pick the best one (the one with the maximum gain).

BET ON IT

If the outcomes describe losses and not gains, the logic is the same but the terms are switched around. Here the worst outcomes are the biggest (or maximum) losses and the best of those is the minimum such loss. This decision criterion is called minimax.

Deciding Which Decision Criterion to Apply

The choice you would make using expected values may not coincide with the one you would make using maximin. If you think about your investment problem, you can surely imagine some people who would prefer GROWTH while others would prefer BONDD. Some people are naturally more or less apt to take risks. Some people have deep pockets, while others cannot afford to lose anything.

Finally, in the preceding example the time horizon was one year. But what if you were planning to park your money for 10 or 20 years? Wouldn't that make a difference? For a long-term investment, your preferred strategy might differ from what you would do with only one year to play with.

I introduced this chapter by talking about states of nature, and I told you that game theory can help you establish strategies for making decisions even when certain elements are completely out of your control. Now I hope you see how using the decision criteria of expected values, maximin, and minimax can improve your outcomes in the face of random states of nature. The restaurant owner doesn't exactly know what the customers will order. Similarly, the investor doesn't know which way the market is going to go. But depending on the situation, some decisions are better on the average while other decisions are safer ones. Let's now develop this concept a little more thoroughly.

Decision Theory

One type of "gaming" behavior that takes place quite often in the business world is called revenue management. If you offer a certain product or service, how can you tweak the prices, promotions, or availability in a way that maximizes your profit? The analysis for such a situation is called *decision theory*.

DEFINITION

Decision theory refers to a set of methodologies based on expected values, maximin, and related criteria that are used to select the best alternative when a decision maker is faced with uncertainty.

Strategic Choices

Suppose you are an airline executive tasked with the job of analyzing a particular flight route. The flights are frequently sold out, but despite this good news, you are still determined to generate some additional profit. Looking at the data, you discover that quite often a small proportion of the passengers who purchase tickets do not actually show up for their flights. You could speculate on why they don't get on board, but rather than engage in armchair philosophy you try to capitalize on their absence.

> **NO BLUFFING**
>
> Lots of companies are catching on to the idea that they can act strategically to improve their revenue streams and cut their costs. Much of this analysis amounts to seeing what they can get away with. For example, a supermarket might not go to any great lengths to restock cans of generic peas if they run out, but they know if they don't quickly restock a big ticket item like baby formula, they'll get burned when hysterical parents end up taking their business elsewhere. Many retailers adjust their prices so customers get a great deal on certain items but then impulsively grab other, high-margin items. The ideas are not necessarily new, but the data mining that helps drive the "gaming" analysis is.

For every passenger who buys a ticket but doesn't board the plane, a seat is freed up! And this means that even if the plane is full, you could have sold one more ticket. The thing is, you never know for sure how many passengers, if any, won't make it on board. The risk involved in overbooking—selling more tickets than seats—is that everyone might show up and then you have to pay through the nose to put the extra passengers on other flights, or put them up in hotels, and so on.

Do you overbook or not? You will only pursue this strategy if it clearly leads to additional profit for your airline. It's time to do some analysis.

The Math of Decisions

Let's assume we know how much additional revenue we get from selling extra tickets, and also how much overbooking will cost when we have more passengers than seats. To keep things manageable, let's suppose that on all full flights, either everyone shows up, or just one or two passengers do not board. Our possible overbooking strategies are as follows:

- Not to overbook any seats
- To overbook by one seat
- To overbook by two seats

Note that overbooking by three or more seats would not be advisable. The reason is that your data indicates that you never have more than two no-shows. If you sell three extra tickets, there will never be a seat available for one of those extra customers. You will have to compensate this passenger every time, and doing so will result in a loss.

The nine possible outcomes are listed in the following table as net profits, and the probabilities regarding how many no-shows you might have are given in the bottom row. (These different possibilities are the various states of nature you have to contend with.)

		Number of Passenger No-Shows		
		0	1	2
Overbooking Strategies	Do Not Overbook	0	0	0
	Overbook by 1	-75	200	200
	Overbook by 2	-150	125	400
Probabilities of No-Shows		0.25	0.40	0.35

Overbooking strategies profits and losses.

Using the Expected Value Decision Criterion

What are the expected values given the different strategies?

If you do not overbook any seats, the outcomes, and therefore the expected value, is zero. Nothing ventured, nothing gained.

If you overbook by one seat, the expected value (EV) is as follows:

$$EV[\text{over by 1}] = (-75 * 0.25) + (200 * 0.40) + (200 * 0.35) = 131.25.$$

If you overbook by two seats, you get the following EV:

$$EV[\text{over by 2}] = (-150 * 0.25) + (125 * 0.40) + (400 * 0.35) = 152.50.$$

Using expected values, the best decision is to overbook by two seats.

Using the Maximin Decision Criterion

What if you employ the maximin decision criterion? You have three choices:

- If you do not overbook, you never lose anything.
- If you overbook by one seat, the worst case is when everyone shows up and you lose $75.
- If you overbook by two seats you might lose $150.

The maximin solution is to find the strategy that maximizes the worst-case outcomes of 0, -$75, and -$150. Since losing nothing is better than losing $75 or $150, the maximin strategy is not to overbook.

> **BET ON IT**
>
> Often the best choice using expected values will conflict with the maximin recommendation. This is not surprising: maximin is cautious while the expected value approach is aggressive. Tailor your approach to the problem at hand. Often, asking yourself what the goal is, or whether you will regret carrying out a certain strategy, is the key to determining which decision criterion to apply.

In the Real World

What do you do if you're the airline executive? In the real world, airlines overbook flights all the time. The executives must be considering the decision using expected values rather than maximin. Even though the airlines get nailed some of the time (when all the passengers show up), they stand to profit whenever there is a no-show, and no-shows must occur frequently enough to pursue the overbooking strategy.

Decision Trees

A decision tree is a branching diagram that takes states of nature into account. Decision trees are similar to the extensive form game diagrams that I introduced in Chapter 1. The primary difference is that where extensive form game diagrams consider different possible strategies for another player, decision trees focus on states of nature.

To illustrate the decision tree approach, let's consider a product liability lawsuit. There is a plaintiff P, who claims that defendant D's product caused serious injury. You will play the role of D's attorney. The relevant facts continue in the following sections.

Creating the Decision Tree

For starters, discussions with P's attorney have indicated that the case can be settled with a payment to *P* of $1.2 million. If your client D goes to trial instead, you believe (from your knowledge of similar cases) that D can expect to pay anything from $0 to $4 million.

There are a number of different possible scenarios. Rather than list them all, I've summarized them in the following decision tree. (Assume that you have some data from similar court cases and can thus estimate the probabilities and award amounts below.)

To denote a point in time where you must make a decision, a square node is used. The circle nodes and their branches represent states of nature (the events that are outside your control). These states of nature have to do with the jury: whether it decides in the defense's favor or else makes you pay up.

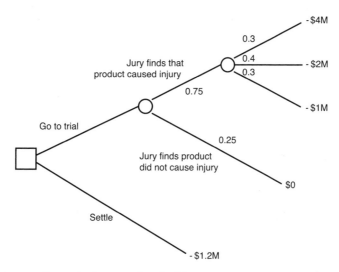

A decision tree for analyzing a product liability case.
(This case is adapted from www.litigationrisk.com/Reading%20a%20Tree.pdf.)

Play It Safe or Roll the Dice?

The question is, should you recommend that the defendant settle and fork over $1.2 million, or should he take his chances in the courtroom? Let's put your game theory tools to work.

If you go to court, either the jury rules in D's favor (whew!) or they find him at fault (bummer!). To figure out the damage, work backward: assume that the case is already in the penalty phase. The expected value at that point is as follows:

$$\text{EV[penalty]} = (-\$4\text{M} * 0.3) + (-\$2\text{M} * 0.4) + (-\$1\text{M} * 0.3) = -\$2.3\text{M}.$$

In other words, if the jury had already found the product at fault, the average payout is $2.3 million. Of course, there is a 25 percent chance that the jury is on the defendant's side and he won't have to pay anything. So altogether, the average payout is:

$$\text{EV[payout]} = (\$2.3M * 0.75) + (0 * 0.25) = \$1.725M.$$

You just figured out that the gamble involved in going to trial is, on the average, more costly than the settlement payment of $1.2 million.

Now let's briefly do a minimax analysis. (Note that since you are dealing with losses, the appropriate method is minimax rather than maximin.)

If you settle you pay $1.2 million for sure. If you go to trial, and the jury rules against your client, the worst outcome is the $4 million penalty. Now, it is true that the jury might rule in your favor. But the conservative logic of minimax tells you that you can't count on a sympathetic jury; instead, you have to assume they are going to hammer you.

Summing up, if you go to trial, on average your client will pay out $1.725 million, and moreover, while you might get lucky in the courtroom, you can't guarantee anything better than a possible $4 million loss going that route. Given this bad news, you and your client might be ready to swallow hard and offer the $1.2 million settlement to make the lawsuit go away.

BET ON IT

The analysis here and throughout the book is not the last word in every situation. For example, after doing the decision tree calculations you wouldn't immediately pick up the phone and surrender $1.2 million to the plaintiff. Instead, you would probably use this figure as a starting point in your settlement negotiations.

Don't give up just yet. You will see some evidence in Chapter 18 that the calculation you just made is not the end of the story. You have not yet been "playing" against anyone. What if you now consider an adversary, for example, the plaintiff's attorneys? It is in fact quite possible that you can get together with the plaintiff's lawyers, all make the same calculations face to face, and yet feel very differently about the results. For now, though, just let the cold, hard numbers guide you.

Statistics, Data, and Trustworthy Conclusions

The courtroom example might leave you wondering how many convicted defendants in real courtrooms are truly innocent, and how many acquitted defendants are actually guilty. Juries can make mistakes! Whether you're considering courtroom verdicts or the results, for example, of medical tests, you really should consider how trustworthy they are.

Is That Swab Reliable?

It would be very hard to discover what proportion of defendants in actual courts are "false positives" (they were found guilty but were truly innocent) or "false negatives" (found innocent but were actually guilty). But very often we can discover such information about medical tests. The result of a lab test may not match the result when a sample is examined a few days later. In fact, for many years, most of the positive results from HIV tests were actually false positives! The following technique, called Bayes' rule, will show you how to reassess the probability of a particular event in the light of new information.

Suppose you have an initial belief regarding the likelihood of an event. For example, in a courtroom the presence of circumstantial evidence may lead you to believe that a defendant is probably guilty. Suppose you quantify this by stating that you believe there is a 75 percent chance the defendant is guilty.

Now suppose you obtain further information, for example, testimony from a certain witness or a particular DNA result. Bayes' rule is a mathematical formula that "updates" your belief by taking into account the additional information. Understanding this Bayesian procedure will strengthen your grasp of the role of updated information in strategic behavior.

Here's an example. Suppose a patient has a sore throat and goes to the doctor. The doctor's office would be able to discover, if they tracked such patients over time, how many patients complaining of sore throats had strep and how many didn't. Suppose that 40 percent of such patients with a sore throat actually had strep throat (and therefore 60 percent did not).

A patient walks in and tells his physician he has a sore throat, and before the doctor asks him to say "aahh," the patient asks her point blank whether he has strep. The doctor says, "Let's do a swab to find out." In the back of her mind, though, the doctor believes that there is a 40 percent chance the patient has strep, based on past information. This 40 percent is called a *prior probability*.

> **DEFINITION**
>
> A **prior probability** is an initial belief about something, often based on historical data.

The physician takes a swab of the patient's throat. Ten minutes later, she will tell the patient whether he has strep throat. But these sorts of tests are not perfect. Some patients with strep will test negative, and some who do not have strep will test positive.

How good are these tests? Suppose 99 percent of those with strep test positive (for the record, this is called the sensitivity of the test), while 98 percent of those who do not have strep test negative (this is called the specificity of the test). Those high percentages give us confidence, at least superficially, in the results.

The questions you want to answer are, given a positive test result, how likely is it the patient has strep, and, given a negative test result, can the physician absolutely rule out strep?

To get these answers, you can do another tree diagram. And to make the job easier, let's run this test for 1,000 sore-throated patients. The following figure makes it a snap.

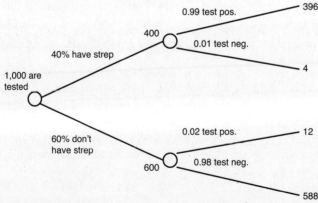

Testing for strep throat.

Let's get this straight. One thousand patients get tested. The prior probability is that 40 percent of them have strep, which amounts to 400 of the patients. When these 400 patients get tested, the test stats are that 99 percent of them test positively. That's 396 out of the 400 people. But bear in mind that 1 percent, or 4 of the 400, will test negative and, therefore, their strep goes undetected.

The other 600 patients do not have strep throat. Ninety-eight percent of them, or 588 patients, will receive the correct test result. But 2 percent of the 600, or 12 patients, will receive a positive test result (a "false positive").

Suppose a patient gets a positive test result. Does he really have strep? Out of all 1,000 patients, 396 of the 400 in the top half of the diagram tested positive, and also 12 of the 600 on the bottom tested positive! So the probability that someone has strep, given that they tested positive, is shown here:

$$396 / (396 + 12) = {}^{396}/_{408}, \text{ which is } 97\%.$$

Conversely, what if the quick swab is negative? Out of the 1,000 patients, 588 in the bottom group tested negative, as did 4 in the top group. The probability that a patient does not have strep, given a negative test result, is as follows:

$$588 / (588 + 4) = {}^{588}/_{592}, \text{ which is over } 99\%.$$

In other words, there is a less than 1 percent chance that a patient with strep goes undetected. Sounds like a pretty acceptable test!

Reliable Tests and Unreliable Results

So what about those HIV tests I mentioned earlier? You might still be wondering how it could be that at one time most of the positive HIV results were wrong!

Suppose everyone in the country was tested for HIV. In most of the developed world the HIV infection rate in the general population is about 1 percent.

Now let's assume that the old HIV tests had the same accuracy—the same sensitivity and specificity—as the strep test. The following diagram is similar to the one for strep, but with a prior probability of 1 percent rather than 40 percent. Also, to make the results clearer, I've changed the population to 100,000 people.

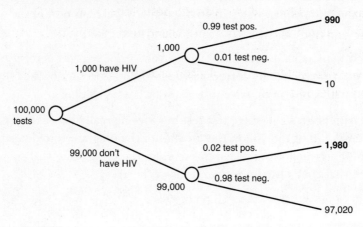

Testing for HIV.

Of the 1 percent (1,000 people) who actually had HIV, 990 of them received a positive test result. But 2 percent of the rest of the population also got a positive result, which amounts to 1,980 false positives. Given these results, the probability that someone with a positive test result did not have HIV is as follows:

1,980 / (990 + 1,980), which is exactly ⅔.

This means that 2 out of every 3 people tested for HIV received a false positive result! So now you have the numbers to back up the old saying that you can't believe everything that you hear, even if it seems to be scientific.

Bayes' rule is a valuable statistical tool, and it plays an important role in much of game theory. Players in games often develop probabilities associated with their opponent's preferences. Then, actual play is observed and used to reassess these probabilities. Games in which players update a prior probability subject to additional information are called Bayesian games.

At this point you have many of the tools you need to solve some important games. What's been missing thus far is a wily and willful opponent, but you are now ready for that kind of struggle.

The Least You Need to Know

- Simple averages are different from weighted averages.
- Maximin decision-making takes the best of the worst scenarios.
- Decision trees map out decision-making with multiple decisions and states of nature.
- Bayes' rule enables you to calculate probabilities once you have some information under your belt.

Zero Sum Games

Chapter 3

In This Chapter

- Making sense of zero-sumness
- Understanding mixed strategies
- Using dominant strategies
- Formulating equilibrium solutions

As you learned in Chapter 1, when journalists and pundits call something a "zero sum game," they are saying that a certain situation is pure conflict: one person's gain is the other's loss.

For game theorists, however, zero sum literally means that in every possible outcome of the game, the payoffs to the two players add up to zero. In other words, if one player has a positive payoff, the opponent must have an equal, but negative, payoff.

There is a certain constancy to real-life zero sum situations that goes beyond mere "winning" and "losing." In this chapter you'll see that when players go at it, "zero sum" and "constant sum" mean essentially the same thing: there is a fixed pie—whatever that "pie" really represents—that is fought over. In this chapter you'll learn the best strategies for these strictly competitive situations.

Set Up and Solution

Think back to the game of Two Finger Morra introduced in Chapter 1. Recall that the players simultaneously put out one or two fingers each. The player who arbitrarily calls "odds" wins when the sum of the fingers is odd; the player who has "evens" wins when the sum of the fingers is even.

> **NO BLUFFING**
>
> This chapter focuses on games with exactly two players, although for storytelling's sake the "players" might be teams or companies. Limiting our examples to two players is not for convenience or to provide for more accessible analysis. It is because when more than two players are involved, there is the possibility that some of the players will join forces—called coalitions—against one or more of the others, and for now we want to focus on the purely competitive nature of zero sum games.

Mixed Strategies

Imagine for a moment that you are going to play Two Finger Morra over and over again with the same friend. You are correct to realize that it doesn't matter how many fingers you put out in any one game. So suppose that in your first game you put out one finger. And suppose time after time you keep putting out just one finger. Surely, after a few games, your friend will notice your pattern and will adjust her play accordingly.

In fact, even little kids realize that they can't follow the same strategy every time. They have to mix things up in order to keep the other kids on their toes. The question is, how do you decide what mix to carry out? In Two Finger Morra, it's easy to imagine implementing this idea of mixing up your strategies: in any one game, you will randomly throw out one finger with a certain probability, and throw out two fingers the rest of the time. This notion is called a *mixed strategy*.

> **DEFINITION**
>
> When there are two strategies to choose from, a **mixed strategy** occurs when one strategy will be played x percent of the time, and the other strategy (100 – x) percent of the time. If there are more than two strategies to choose from, a mixed strategy is found when the probabilities assigned to the various strategies all add up to 1 (i.e., 100 percent).
>
> A **pure strategy** is when a player selects only a single action (or a particular sequence of actions) from a set of alternatives.

Pure Strategies and Domination

Surprisingly, there are some games where the solution does not involve mixing strategies at all. You know you are going to play a certain strategy, and your opponent knows it, too; and yet, there's no point switching to something else! When every player will use a single strategy, game theorists say that the game has a solution in *pure strategies*. Let's look at an example.

In the zero sum game represented in the following table, Player 1 chooses the row while Player 2 selects the column. Each of the possible strategy pairs results in an outcome listed in the table that shows Player 1's payoff followed by Player 2's payoff. If you like, you can think of units of money changing hands, but let's call them "points." Either way, what one player gains, the other loses. Each pair of payoffs adds up to zero.

		Player 2	
		Column 1	Column 2
Player 1	Row 1	2, -2	-3, 3
	Row 2	5, -5	1, -1

A zero sum game with a solution in pure strategies.

All the outcome pairs in the table add up to zero. For example, if Player 1 picks row 1 and Player 2 picks column 1, then Player 1 wins two points and Player 2 loses two points.

Notice that Player 1 comes out better in three of the four outcomes, but that in row 1, column 2, Player 2 wins three points while Player 1 loses three.

So how can the players do their best? Suppose Player 1 is considering row 2. His thinking is this: "Suppose Player 2 picks column 1. I am better off in row 2 because the row 2 payoff of five points beats my row 1 payoff of two points."

And Player 1 further realizes, "If Player 2 picks column 2 instead, I'm still better off in row 2 than in row 1, because winning one point in row 2 is better than losing three in row 1."

In other words, row 2 is better for Player 1 no matter what Player 2 does! And the bizarre thing is, Player 1 can even send Player 2 a text message before they play, declaring his intention to play row 2 and this revelation won't ruin it for him. A strategy like row 2, which, no matter what, is always at least as good for a player as another strategy, is said to dominate the other strategy. Game theorists say that strategy A dominates strategy B for a certain player if every outcome for that player under strategy A is better than the corresponding outcome under strategy B.

Now let's focus on Player 2. Column 2 for her dominates column 1. If Player 1 plays row 1, then she will prefer column 2 because winning three points is superior to losing two (the outcome if she had played column 1). And if Player 1 were to play row 2, Player 2 is still better off in column 2 because losing one point is preferable to losing five points (had she played column 1). No matter what Player 1 does, Player 2 is better off in column 2.

So row 2 dominates row 1, and column 2 dominates column 1. This leaves the two players in the lower right-hand corner, with the outcome (1,-1). This means that Player 2 will give up one point to Player 1. Furthermore, if either player tries to fool the other by playing a different strategy, the player who switches will come out worse than before!

The Equilibrium as a Solution

The preceding exercise introduced a game in which the best strategies were pure and not mixed. It also introduced the most important solution concept in game theory: the notion of an *equilibrium*.

If you look up "equilibrium" in a dictionary you'll find that it refers to a state of balance, perhaps a state of rest. That's what we have here. Neither player has any incentive to change his or her strategy—there's no way to improve the results—so each stays put at the outcome (1,-1). And even though Player 2 is not thrilled with this state of affairs (she loses one point every time), it's the best she can do.

DEFINITION

An **equilibrium** in a zero sum game is a pair of strategies such that neither player will improve their payoff by (unilaterally) deviating from his or her strategy.

Sometimes equilibrium analysis requires us to get into the other player's head so as to anticipate their best strategy. Consider the game represented in the following table.

	Player 2	
Player 1	2, -2	-1, 1
	4, -4	-3, 3

A zero sum game with a single dominant strategy.

When Player 1 looks for row domination, he won't find any: if Player 2 plays the first column, Player 1 is better off in the second row. But if Player 2 plays the second column, Player 1 is better off in the first row. Since neither row dominates the other, it is not obvious at this point what Player 1 should do.

But what if Player 1 imagines himself in Player 2's position? Looking at the game as the column player, Player 1 observes that the second column dominates the first column. If Player 1 plays the first row, the payoff of one point for Player 2 in column 2 is better than the payoff of losing two in column 1. And if Player 1 plays the second row, the column 2 payoff of three points for Player 2 sure beats the loss of four points if Player 2 chooses the first column instead!

So now Player 1—understanding how Player 2 will be thinking—can expect Player 2 to play column 2 as a pure strategy. Therefore Player 1 realizes that he will be stuck with the column 2 payoffs. To make the best of it, Player 1 will play the first row. (Losing one point in row 1 is better than losing three in row 2.)

Player 2 will do the same analysis. She will grasp that column 2 is always better for her, and further that the best thing for Player 1 is to play the first row.

Finally, notice that the (-1,1) outcome is an equilibrium of this game. Neither player can deviate from the prescribed strategies without doing worse, so there is no chance of successfully outfoxing the other player.

War Games

Sometimes pure strategy equilibria end up being solutions in real life situations. I won't discuss many military examples in this book, but even if your knowledge of military planning begins and ends with the goofy image of Peter Sellers in *Dr. Strangelove*, you can certainly understand why military strategists around the globe are interested in the very same thought processes that game theorists are.

Let's look at an excellent example of a pure strategy equilibrium that occurred in the Battle of the Bismarck Sea, a South Pacific World War II naval battle that took place in February 1943.

Briefly, the situation was this: the Japanese commander needed to send a convoy to deliver reinforcements and supplies to Japan's base on the island of New Guinea. He had the choice of sending the convoy on a northern or southern route around the neighboring island of New Britain. The northern route was somewhat protected from attack due to rainy weather and poor visibility.

Either route would take the convoy three days. The American commander had orders to intercept and inflict maximum damage on the convoy. Essentially this war game amounted to the Japanese selecting between the northern or southern route, and the Americans guessing which route to attack. If the Americans made the wrong choice, they would lose valuable time doubling back.

The following matrix shows the available strategies for both commanders. The payoffs are described as numbers of days of ensuing bombing. (The more days of bombing, the better for the Americans and the worse for the Japanese.)

Instead of assigning a numerical value to the outcomes, battle scenarios are described instead. It is important to note that the analysis is equally valid even though the bombing period isn't converted into utilities.

		Japanese Choices	
		Sail North	Sail South
American Choices	Attack North	2 days	2 days
	Attack South	1 day	3 days

Expected number of days of bombing.

> **BET ON IT**
>
> It turns out that matrix games can sometimes be analyzed even when the payoffs are not expressed as purely numerical values. If there is a pure strategy equilibrium, you can find it in the table. However, if there is no pure strategy equilibrium, you would have to express the payoffs purely numerically (not as days of bombing, etc.) in order to find a mixed strategy equilibrium.

For the American side, neither strategy dominates the other. But for the Japanese, the northern route dominates the southern route. (If the Americans guessed northern, the outcomes for either Japanese strategy are the same, but if the Americans went south, then the Japanese would be much better off going north.)

So the Japanese commander ended up using his dominant strategy, taking the northern route. The American commander realized how his counterpart would have to sail, and he sent his bombers north as well. The outcome was grave for the Japanese, but as we can see from the table, there was nothing better to be done.

Constant Sum vs. Zero Sum

In Chapter 1 I briefly mentioned a television network programming game. There were two TV networks, where each network was to choose one of two programs that would run opposite one another during prime time.

Let's assume that there is a certain viewership that will be watching either a program on Network A or one on Network B. This audience will be divided into some percentage that will be watching Network A (NBC), while the remaining percentage of viewers will be watching Network B (CBS).

NBC and CBS have each narrowed down their choices to running either a sitcom or a sports program. As seen in the table, there are four possible outcomes. Suppose the networks each have access to research that gives the market shares in the four cases. These market shares, given as percentages of the audience, are listed in the following table.

		CBS	
		CBS sitcom	CBS sports
NBC	NBC sitcom	56%, 44%	47%, 53%
	NBC sports	46%, 54%	48%, 52%

Constant sum market share game.

The market share percentages for each outcome add up to 100 percent, making this a constant sum game.

Constant sum games certainly seem to behave like zero sum games because in each case a gain for one player is balanced by a loss for the other. But aside from this obvious similarity, can we say that they are equivalent?

BET ON IT

Every constant sum game can be transformed to a zero sum game without changing the nature of the contest.

The answer is yes, and in fact, you can take any constant sum game and convert it to an equivalent zero sum game.

Consider any outcome pair in the constant sum game. Let's call this pair (u_1, u_2). Since the game is constant sum, every such outcome pair adds up to the same constant. Let c represent this constant, which means that $u_1 + u_2 = c$.

We will transform the outcome (u_1, u_2) into an outcome (v_1, v_2) by the following trick:

Let $v_1 = 2u_1 - c$, and let $v_2 = 2u_2 - c$.

(If you'd like, you can do a little algebra to convince yourself that $v_1 + v_2 = 0$, meaning that the new outcomes are actually zero sum.)

In our market share example, the constant c is equal to 100. Let's do the math: the outcome in row 1, column 1 is (56, 44), in percent. So let $u_1 = 56$ and $u_2 = 44$.

Then $v_1 = (2 * 56) - 100 = 12$, and $v_2 = (2 * 44) - 100 = -12$.

Wow, it worked! Try out the remaining three outcomes. You should end up with the following matrix.

		CBS	
		CBS sitcom	CBS sports
NBC	NBC sitcom	12, -12	-6, 6
	NBC sports	-8, 8	-4, 4

Zero sum market share game.

The pairs in this table represent the numbers of percentage points the networks are ahead or behind. For example, when both run a sports program, the original figures were (48%, 52%), so the (-4, 4) entry in this table correctly shows that NBC trails CBS by 4 percentage points, and equivalently that CBS is ahead by the same 4 points. As you can see, the new zero sum game really does represent the original situation.

Solving Zero Sum Games

It's time to move up to the big leagues: to solve zero sum games in general. Most such games do not have equilibrium solutions in pure strategies, so we need to come up with the right mixed strategies.

Mixed Strategy Equilibria

Before we go any further, scan the converted zero sum network game for dominant strategies. If you didn't find any, that's good, because no strategy is dominant. This means that both players will have to resort to some sort of mixing, or random picking, of their strategies. Before working through the solution mathematically, here are the key steps to the analysis:

1. Each player figures out what their average payoff is under each of the other player's pure strategies and writes that down.
2. Each player then sets those average payoffs equal to one another and solves for the appropriate strategy mix.

This guarantees a certain payoff no matter what the other player does!

BET ON IT

When you equate your expected payoffs resulting from any of the opponent's strategies, you obtain a guaranteed payoff that your opponent can't mess with.

Solve the Problem for NBC

Let's first solve the problem for NBC. There is no pure strategy equilibrium, so NBC will have to play a mixed strategy in which they will run a sitcom x percent of the time and run a sports program the rest of the time. From now on we will drop the percentages and use x and $(1 - x)$ to denote the percentages as probabilities.

Do the Math

Follow these steps:

1. Write down the expected payoff to NBC under each of CBS's two pure strategies.
2. Equate the two expected payoffs and solve for the value of x.

Why equate the two payoffs? Suppose the two payoffs were different, for a certain value of x. And suppose that NBC is better off when CBS runs its sitcom as opposed to the sports program. That's good for NBC, right? But the CBS strategists aren't clueless—they will realize this and run the sports program instead, to its advantage (and NBC's disadvantage).

So we equate the two expected values to "immunize" NBC against either of CBS's strategies. By doing this, NBC guarantees itself a certain market share no matter what CBS does. This is exactly the maximin approach from Chapter 2.

Payoffs to NBC Under Each of CBS's Strategies

	CBS picks sitcom		CBS picks sports
	$12 * x + (-8) * (1 - x)$	$=$	$(-6) * x + (-4) * (1 - x)$
So	$20 * x - 8$	$=$	$-2 * x - 4$
and we get	$22 * x$	$=$	4
which means that	x	$=$	$2/11$

The equilibrium mix for NBC is to play row 1 $2/11$ of the time, and row 2 the remaining $9/11$ of the time. This literally means that NBC will randomly pick the sitcom $2/11$, or about 18 percent, of the time, and randomly pick the sports program the other 82 percent of the time.

The Analysis

At first blush it might seem odd that game theory would have you make an important executive decision—selecting a prime time TV program—by essentially flipping some kind of coin (albeit an $18/82$, not a $50/50$, coin). But it's the exact same idea as when you played Two Finger Morra as a kid. Every time you were about to put out one or two fingers, you had to scramble your mind and try to pick randomly.

You might also be surprised that NBC will be picking its sitcom only about 18 percent of the time. The sitcom seemed to be NBC's stronger program. But look at it this way: CBS will want to avoid running into NBC's sitcom with their own sitcom, which means that CBS will run its sports program most of the time anyway. (And then NBC's sitcom does badly, but their sports program improves on their sitcom!)

Finally, you might wonder where this $18/82$ mix leaves you. Are you gaining or losing market share overall? I suspect you think the average outcome isn't too good. Let's figure it out—it's easy! All you need to do is to take either of the above expected value expressions (remember, they're equal) and plug in the value of $2/11$ for x.

Your average payoff is: $12 * (2/11) + (-8) * (9/11) = -48/11$.

This means that on average, NBC will be about 4.36 percentage points behind CBS. Ouch!

POISONED PAWN

Don't look at the correct mixed strategy as advice you can laugh off. If you deviate from the mixed strategy prescription, you will do worse on average.

Solve the Problem for CBS

Now look at the situation from CBS's point of view. Perhaps you have the intuition that CBS is in the driver's seat, but you still need to figure out just how to play.

CBS is going to apply the exact same kind of reasoning that you just observed for NBC. CBS will know that NBC runs either their sitcom or sports program. To guarantee an average payoff, CBS will equate its own expected payoffs under both of NBC's choices.

Do the Math

CBS will run their sitcom y of the time and run their sports program $(1 - y)$ of the time. Now you need to equate CBS's payoffs when NBC picks either sitcom or sports.

Payoffs to CBS Under Each of NBC's Strategies

	NBC picks sitcom		NBC picks sports
	$-12 * y + 6 * (1 - y)$	=	$8 * y + 4 * (1 - y)$
So	$-18 * y + 6$	=	$4 * y + 4$
and we get	$22 * y$	=	2
which means that	y	=	$1/11$

The solution is that CBS will run its sitcom only $1/11$ of the time (about 9 percent) while they will pick the sports program $10/11$ of the time (about 91 percent).

What is the average payoff that CBS is able to guarantee for itself? To find out substitute the value $y = 1/11$ into either of the expected payoff expressions above.

CBS's guaranteed average payoff is:

$-12 * (1/11) + 6 * (10/11) = 48/11$, or about 4.36 percentage points ahead of NBC.

Now wait a minute! Didn't we see the exact same average value when we looked at NBC? Yes we did, and the fact that these values coincide is no coincidence. What we have just found is one of the first and most important results ever discovered in game theory: in zero sum games, the most that one player can guarantee winning is equal to the least that the other player can guarantee losing.

> **BET ON IT**
>
> In 1928, John von Neumann proved the so-called minimax theorem: that for zero sum games, the maximin outcome for the row player is always equal to the minimax outcome for the column player. In doing this he showed that every zero sum game has an equilibrium that is determined by these strategies.

What's really helpful—and cool, too—is that you can draw a graph that superbly illustrates what's going on. Let's draw the outcomes from NBC's point of view. The following graph has two vertical axes. The axis on the left shows the two possible payoffs to NBC when $x = 0$ (i.e., when NBC runs the sports program for sure); the axis on the right shows the two possible payoffs to NBC when $x = 1$ (i.e., NBC runs the sitcom).

Look at the two crossing lines. They correspond to the two pure strategies for CBS. Suppose CBS goes with the sitcom. If NBC runs sports, the payoff for NBC is -8, while if NBC runs its sitcom, NBC's payoff is 12.

Now here's the key: when NBC mixes between the two programs, their average payoff is given along the line between the -8 and the 12. When CBS runs its sports program, NBC's payoffs vary between -4 and -6, as shown by the other crossing line in the graph.

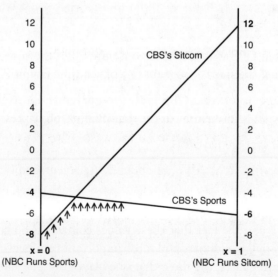

Payoffs to NBC when CBS runs sitcom or sports.

Where the lines intersect is the mixed strategy ($2/11$, $9/11$). That is, the point of intersection is $2/11$ of the way from the left axis to the right axis.

Finally, what's with those little arrows in the graph? The idea of the arrows is to show, for various mixed strategies, what the worst-case payoff is for NBC. The worst-case payoffs run along the "lower envelope" that is indicated by the arrows. The best of these worst-case payoffs is the highest such point—found at the intersection of the lines.

The graph helps illustrate that deviating from the correct mixed strategy will hurt you. If NBC runs the sports program more than 82 percent of the time, they are getting too close to the vertical $x = 0$ axis, and CBS will punish them by running its sitcom. Look at the graph: to the left of the intersection point, NBC's payoffs go down when CBS runs its sitcom!

Moreover, if NBC runs its sitcom more than 18 percent of the time, CBS will punish them by running its sports program. Again, in the graph, to the right of the intersection point, NBC's payoffs decrease when CBS runs its sitcom. Another way to express this is that the intersection point provides NBC its best possible guarantee—literally the maximin payoff.

These observations give us one more important piece of the puzzle. The two mixed strategy pairs you just calculated—NBC plays an $18/82$ mix between sitcom and sports while CBS plays a $9/91$ mix between sitcom and sports—satisfy the following two important game conditions:

- NBC plays maximin while CBS plays minimax.
- The two mixed strategies are in equilibrium.

Remember, being in equilibrium means that if either player deviates from the prescribed strategy—mixed strategies in this case—that player will not improve their payoff.

NO BLUFFING

In Chapter 1 I mentioned that von Neumann was inspired by the idea of improving his poker play. As it turns out, poker is too complicated for game theory to be a useful guide.

Finally, one valuable way to think about this market share encounter is as a repeated game. Even though our market share game is technically a one-and-done situation, imagining the game repeated over time helps to conceptualize it. The "best play" will have the two players randomly picking sitcom or sports on each round of the game, and over time you will see NBC's sitcom being run about 18 percent of the time, CBS's sports program running about 91 percent of the time, and so on.

Saddle Up

A saddle point is a term mathematicians use to describe coordinates that are at the bottom of a dip in one direction, but at the top of a hill in the transverse direction. This happens sometimes in zero sum games, when the maximin outcome for the row player using a pure strategy coincides with the minimax solution, under a pure strategy, for the column player.

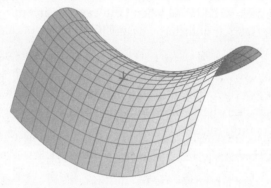

A saddle point.

Here's a whimsical example: two buddies are planning to camp at a national park. The campsites are located at intersections of east-west trails with north-south trails. Buddy A likes high altitudes; Buddy B likes low altitudes.

To decide on a site, Buddy A will pick an east-west trail, while Buddy B will pick a north-south trail. The altitudes of the campgrounds (in thousands of feet) are listed in the following table. In this table there's just one payoff (the altitude) for each outcome. The game is still zero sum in nature because the pleasure that Buddy A receives from each foot of elevation is equally matched by Buddy B's discomfort.

		N-S		
		Col. 1	Col. 2	Col. 3
	Row 1	7	2	1
E-W	Row 2	2	2	5
	Row 3	5	3	4

Trail intersections and their altitudes (thousands of feet).

Buddy A's maximin strategy is to take the best (highest) of the worst row payoffs. We first identify Buddy A's worst payoff for each of his options:

- His worst, or lowest, payoff in row 1 is 1
- His worst payoff in row 2 is 2
- His worst payoff in row 3 is 3

Now Buddy A takes the maximum (highest) of those three worst-case scenarios. This means that Buddy A will pick row 3, and will do no worse than the 3,000 foot campground in that row.

Buddy B's minimax strategy is to take the best (lowest) of the worst column payoffs. The worst column payoffs for B are the highest altitude ones: 7 in the first column, 3 in the second column, and 5 in the third column. So Buddy B will pick column 2, and do no worse than the 3,000 foot high campground there.

Since they are both going for the same campground (row 3, column 2), we have a saddle point. Notice how we have arrived at an equilibrium. If either player unilaterally deviates from this strategy pair, he will end up worse off!

Supersized Strategy Sets

Now it is time to come clean. Maybe when you were going along with the sitcom versus sports example, you wondered how a player might be able to add another option to the mix. (Indeed, in 2009 NBC decided to break the traditional network programming mold by putting late-night talk show host Jay Leno into a prime-time slot.) Or, looking at the saddle point example, you might have seen that the matrix showed three choices for each player and wondered how such a game could be solved if it had no saddle point.

Indeed, one of the most well-known zero sum games is Rock, Paper, Scissors. (Rock smashes scissors, scissors cut paper, but paper covers rock.) This game has three possible strategies for each player and no pure strategy equilibrium point. How do you solve it?

> **BET ON IT**
> Many games have more than two strategies for each player and can't be solved as easily as the two-by-two examples here.

First, the quick answer: the game has a special symmetry to it, and it turns out that the equilibrium solution reflects that symmetry. The mixed strategy equilibrium is for both players to randomly choose rock, paper, or scissors a third of the time each.

The more general answer to solving zero sum games with more than two rows and columns is that they are solved using computer programs. You can use a mathematical technique called "linear programming" to find the mixed strategy equilibrium points in larger games.

Misconceptions About the Zero Sum Concept

Before moving on, it's important that you understand what zero sum games are not. If you are a soccer or hockey fan, then beware that despite the games being won, lost, or tied on the field, the way the points are allocated in the standings, and therefore the way the teams are ranked, is absolutely not zero sum.

As I write this, the Los Angeles Galaxy and the Houston Dynamo are tied in the North American Major League Soccer Western standings. Each team has played 30 games. Here are their records:

	Games	W	L	T	Points
Los Angeles	30	12	6	12	48
Houston	30	13	8	9	48

Soccer rankings in North America (and football rankings elsewhere; hockey is a little different) award three points for each victory, one point for each tie, and zero points for each loss. So despite the fact that you might believe LA ought to be ranked higher than Houston because their 12-6 won-lost record is "better" than Houston's 13-8 record, league officials do not see it that way.

Even though the result of a soccer match is a W, L, or T, the numerical outcomes are not zero sum. More precisely, they are not constant sum. Here's why. Suppose the Galaxy and Dynamo play each other. If one of the teams wins, that team receives three points (the other one receives zero points) and the sum between the teams is 3. But if the two teams tie, the sum between them is 2. Something here literally doesn't add up!

> **POISONED PAWN**
>
> Beware that just because the game result is called a win or loss does not necessarily mean that the situation is truly zero sum.

What's going on is that the soccer leagues have created this very imbalance in order to produce a more exciting entertainment product. If the Galaxy and Dynamo play each other twice, and they play defensively (that is, not to lose), the games might both end in a tie, each team would accumulate two points in the league table, and many fans would yawn.

But if they both play aggressively, they will create more scoring opportunities on each side. Attacking play will probably lead to more victories—and defeats. Suppose the Galaxy and Dynamo play two games, and win one game apiece. Then each comes away with a total of three points, better for both than if they had two ties. By the way, if soccer leagues counted wins as one point, ties as half a point, and losses as zero, the points, which now add to one for any match, would be consistent with the "natural" outcomes of the physical contest.

So, to reiterate, the results being called wins and losses are not congruent with the points that accrue in the standings. In fact, the current points system does three things: first and most directly, it creates an incentive for teams to play more aggressively; second, the by-product of the more aggressive and exciting play is higher ticket sales; and … third? The third point is that the real game played here is not a win-lose, zero sum affair. The real game, as we saw when tallying up the points, is nonzero sum. Let's take that lesson into the next chapter.

The Least You Need to Know

- Zero sum and constant sum games are equivalent in that one player's gain is at the other's expense.
- A mixed strategy is when a player assigns a positive probability to more than one strategy.

- You can't go wrong with a dominant strategy, if one is available.
- A pair of strategies is in equilibrium when neither player can improve their payoff when unilaterally defecting from their strategy.
- Every zero sum game has an equilibrium in mixed strategies (if not in pure strategies).

Nonzero Sum Games

Chapter 4

In This Chapter

- Understanding the benefits of studying nonzero sum games
- Identifying nonzero sum equilibria
- Computing strategy mixes and payoffs
- Eliminating nonsensical equilibria

Our species has been able to create an astonishing amount of stuff in the space of just a few thousand years. But not all of this productivity can be explained by the notion of zero sum and its fixed pie.

Many human interactions benefit from a synergy in which different people bring together complementary resources and skills with the result being *more* than the sum of the parts. Consider a simple market: you have a surplus of wheat and I have a surplus of eggs. Left to our own devices a lot of the wheat and eggs might go to waste. But when we trade, we are both far better off. This is certainly not a zero sum scenario.

It is equally possible, however, that joint actions can leave both decision makers worse off. For example, you and I might both overfish the waters today, only to discover that the fish are all gone tomorrow. In this situation, neither one of us are better off. Nonzero sum games are worth studying because they can capture much of this type of complexity. In this chapter you will learn how the nonzero sum model captures more of the complexity of our interactions than zero sum games do. You will also uncover some challenges that the nonzero solution approach generates.

Nonzero and Its Rich Flavor

The author Robert Wright goes so far as to say that nonzero-sumness is the very force that has brought our planet from isolated pools of bacteria to a network of globally interconnected, democratic, technically sophisticated societies. He argues that what powers our cultural evolution and provides the ratchet in our forward progress is a certain force, a potential, the very essence of which is the notion of nonzero sum. Since this is not a philosophy book, let's come down to Earth and start to pinpoint exactly what "nonzero" means.

The nonzero sum games in this chapter employ the same kind of game matrix as the zero sum games in Chapter 3. There are still just two players in these games. The only difference is that the sum of the numerical outcomes over the strategy pairs varies. In other words, the payoffs do not generally sum to zero or to any fixed number.

It's best to get started with an example, and what better one than the Coordination game that you first had a look at in Chapter 1?

The Coordination Game

The main solution concept in nonzero sum games, is to find an equilibrium, just as we do with zero sum games. But the idea of an equilibrium in nonzero sum games is a little more involved than it is for zero sum games.

Let's review the basic facts in our Coordination game: Sony and Toshiba would like to agree to adopt a common high-definition technology. They will each choose to back either Blu-ray or HD DVD. If they choose the same format, they will adopt it; if they disagree, they will go their separate ways with incompatible formats.

They privately write down their choices and then compare them. Jointly going ahead with Blu-ray is advantageous for Sony; proceeding with HD DVD is better for Toshiba. Therefore the payoffs are lopsided in a particular way. The following table illustrates the situation with scaled outcomes.

		Toshiba's Choices	
		Blu-ray	HD DVD
Sony's Choices	Blu-ray	(4, 1)	(0, 0)
	HD DVD	(0, 0)	(1, 4)

The Coordination game matrix for Sony and Toshiba.

Nash Equilibria

When solving for equilibria in zero sum games you don't worry about the opponent's payoffs, just their strategies. The reason is that the opponent's payoffs are implicitly known—they are always diametrically opposed to yours. But in nonzero sum games, when computing the equilibrium strategies, you have to consider what the opponent's payoffs are. As you might expect, the solution technique will be a little different.

In 1950, John Nash, as a 21-year-old grad student at Princeton, solved the problem of finding equilibria in nonzero sum games. His idea reinterprets the idea of equilibrium strategies as being best replies to the other's strategy.

NO BLUFFING

John Nash wrote a number of brilliant papers on game theory and other branches of mathematics when he was in his twenties, before he began to suffer from schizophrenia. He is the subject of the book, *A Beautiful Mind*, by Sylvia Nasar, which was made into an Oscar-winning movie.

Pure Strategy Nash Equilibria

Take a look at the Coordination game matrix. Imagine that Sony picks Blu-ray, while Toshiba decides to give in and opts for Blu-ray as well. The outcome is much more convenient for Sony but is better for Toshiba than if they didn't coordinate. The payoff of (4, 1) reflects this.

The thing is, (Blu-ray, Blu-ray) is a pure strategy equilibrium. Both players are playing their best reply to the other's strategy. If Sony and Toshiba both pick Blu-ray, they are at the outcome (4, 1). Suppose Toshiba reconsiders and is thinking about picking HD DVD. Given that Sony's choice has restricted the outcomes to the first row of the matrix, Toshiba's choice of HD DVD will leave it with a payoff of 0, which is worse than the payoff of 1 from selecting Blu-ray. So Toshiba's best response is to stick with Blu-ray.

Similarly, (HD DVD, HD DVD) is another pure strategy equilibrium. Game theorists call these pure strategy equilibria in nonzero sum games *Nash equilibria*.

> **DEFINITION**
>
> A **Nash equilibrium** in a two-person, nonzero sum game is when both players are employing their best reply strategies.

Mixed Strategy Nash Equilibria

The Coordination game represents a commonplace problem in the world of business and technology and beyond. And although it's nice to know there are two "reasonable" solutions to the game in the form of the Nash equilibria, neither solution is entirely satisfactory. Either Sony or Toshiba is going to be somewhat inconvenienced. It is important to discover whether a mixed strategy Nash equilibrium exists, and if so, whether it somehow improves on the pure strategy pairs.

Finding the mixed strategy equilibrium is no harder than it was for zero sum games, but the logic of it is more obscure. One comment before getting started: feel free to skip the algebraic solution. Other than missing out on the solution technique, if you move ahead to the solution itself and its analysis, you'll be fine.

To solve for the mixed strategy Nash equilibrium, you solve for each player's mix one at a time. The guiding principle behind finding the best reply is that for any player, each pure strategy that is used in their equilibrium mix must yield the same expected payoff. This means that each player must solve for the mix that equalizes the other player's average payoffs!

The Algebraic Solution

If you're Sony, you will find the mixed strategy x for Blu-ray and $(1 - x)$ for HD DVD that makes Toshiba indifferent between their two choices. We use Toshiba's payoffs and equate its expected values for Blu-ray and HD DVD.

	If Toshiba picks Blu-ray		If Toshiba picks HD DVD
	$1 * x + 0 * (1-x)$	=	$0 * x + 4 * (1-x)$
	x	=	$4 - 4 * x$
	$5 * x$	=	4
So	x	=	$4/5$

So Sony's equilibrium strategy is to randomly pick Blu-ray $4/5$ of the time, and to pick HD DVD the other $1/5$ of the time.

Now Toshiba will mix strategies (y, $1 - y$) to neutralize Sony's expected outcomes under each of Sony's strategies. (We use Sony's payoffs now.)

	If Sony picks Blu-ray		If Sony picks HD DVD
	$4 * y + 0 * (1-y)$	=	$0 * y + 1 * (1-y)$
	$4 * y$	=	$1 - y$
	$5 * y$	=	1
So	y	=	$1/5$

Toshiba is thus advised to pick Blu-ray (randomly) $1/5$ of the time and HD DVD the remaining $4/5$ of the time.

The Results

The Nash equilibrium strategy has both Sony and Toshiba go for their preferred choices $4/5$, or 80 percent, of the time, and give in (randomly) the other $1/5$, or 20 percent, of the time. This might have a certain intuitive appeal. But what are the average payoffs?

Equilibrium Payoffs

Finding the equilibrium solution is crucial but you will also want to determine the average payoffs to the players when they play their equilibrium strategies. Let's do Sony's equilibrium payoff first. (Since the game is symmetric, i.e., Toshiba's strategies and payoffs precisely mirror Sony's, Toshiba's equilibrium mix will be identical.)

Sony picks Blu-ray $4/5$ of the time. But the two players coordinate on only $1/5$ of those selections since that is the probability that Toshiba will also pick Blu-ray. The other $4/5$ of the time they do not coordinate.

Sometimes Sony will give in and pick HD DVD; in fact, this will happen the other $1/5$ of the time. When Sony picks HD DVD, $1/5$ of the time Toshiba gives in and picks Blu-ray (tragic!) and $4/5$ of the time Toshiba picks HD DVD.

Four outcomes are possible, and now you can find the probabilities associated with all of them. For example, the probability that both pick Blu-ray is $4/5 * 1/5$. The following expression gives the expected value for Sony under its equilibrium strategy:

$(4/5) * (1/5) * 4 + (4/5) * (4/5) * 0 + (1/5) * (1/5) * 0 + (1/5) * (4/5) * 1 = 4/5$.

Unfortunately, Sony realizes that this mixed strategy equilibrium leaves them worse off than if they had ended up with HD DVD. Because the game is entirely symmetric, the equilibrium payoff for Toshiba is the same as Sony's, i.e., equal to $4/5$. Thus Toshiba discovers that they would have been better off agreeing to switch to Blu-ray.

BET ON IT

Nash proved that every nonzero sum game has at least one (Nash) equilibrium, possibly in mixed strategies. This might not sound earth-shattering, but imagine if it weren't true. Then we would have no reasonable solution concept at all to rely on! But whether a Nash equilibrium is always a "good" outcome is an entirely different story!

Now you might be thinking, "the heck with game theory, how about we just do a coin flip?" This sounds a lot more straightforward than the $80/20$ and $20/80$ mixed strategy equilibrium, or the pure strategy equilibria where one player or the other feels trapped. Let's check it out!

If Sony and Toshiba each flip a coin, there are four possible outcomes (HH, HT, TH, and TT, where H represents heads and T represents tails) and an equal 1 in 4 chance that they end up in any of them. The average payoff for either player is equal to:

$$(1/4) * 4 + (1/4) * 0 + (1/4) * 0 + (1/4) * 1 = 5/4.$$

An average "utility" of $5/4$ beats the average of $4/5$ from the mixed strategy equilibria, and also beats the payoff of 1 when one player gives in to the other! Just flipping a coin sounds promising. But what if Toshiba knows Sony is going to flip a coin? Toshiba is better off (on the average) picking purely HD DVD! Half of the time Toshiba would get their preferred format (with payoff of 4), and the other half of the time they would remain incompatible (= 0), for an average payoff of 2.

This means that flipping a coin in the Coordination game is not the best reply when the other player flips a coin. The math here is instructive. Toshiba is trying to find the best mix $(y, 1 - y)$ in response to Sony's coin flip. Toshiba's average payoff is as follows:

$$y * (1/2) * 1 + y * (1/2) * 0 + (1-y) * (1/2) * 0 + (1-y) * (1/2) * 4.$$

This average payoff for Toshiba is maximized when $y = 0$. That means Toshiba is never going to pick Blu-ray. Put differently, the math confirms that Toshiba's best response to Sony's coin flip is to always choose HD DVD. And since we assume that

Sony can figure this out, too (rationality being an equal opportunity forte), then Sony will not flip a coin either. So neither player will want to flip a coin, leaving both companies staring at the other solutions once more.

These conclusions might leave you frustrated. But all is not lost. Real life players, for example technology companies, just might find a way to negotiate their way out of such standoffs. I provide some suggestions for doing so in Chapter 5.

> **POISONED PAWN**
>
> One troubling thing to game theorists is that many nonzero sum games have more than one Nash equilibrium. This might sound good, but it isn't! The different equilibria can result in wildly different outcomes. Also, some equilibria are pretty implausible, meaning that it is unlikely they could be implemented in real life. Not knowing which equilibrium to go for can leave players out on a limb.

Tweaking Payoffs and Shifting Strategy

Before moving on from the Coordination game, there is something else to consider. For convenience with respect to numerical outcomes, we treated Toshiba's utility for HD DVD equal to Sony's taste for Blu-ray, and so on. But how could we measure these utilities in real life, and what if they differed? For example, what if Sony would hugely profit from the joint adoption of Blu-ray, while Toshiba would only mildly profit from joint adoption of HD DVD? If we were measuring their utility on a 0 to 10 scale, what if the payoff matrix looked like the following:

		Toshiba's Choices	
		Blu-ray	HD DVD
Sony's Choices	Blu-ray	(10, 1)	(0, 0)
	HD DVD	(0, 0)	(1, 4)

Asymmetric payoffs in the Coordination game.

The only departure from the previous payoff matrix is that Sony's appetite for Blu-ray has gone from 4 to 10. Will this adjustment change the nature of the solutions?

Rather than go through the algebra in detail, let's cut to the chase and provide the solutions. (If you want to do the math, go back to the equation where we initially calculated the mixed strategy equilibrium and substitute the payoff of 10 for Sony when computing Toshiba's equilibrium mix.)

Even if Sony has turned into a total Blu-ray extremist, its equilibrium mix will not change! (Remember, Sony's mix depends only on Toshiba's payoffs.) But Toshiba's equilibrium mix will change. As it turns out, Sony will still play ($4/5$, $1/5$) but now Toshiba's equilibrium mix is ($1/11$, $10/11$). This is fascinating: Toshiba's response to Sony's having become super-hooked on Blu-ray is to pick Blu-ray even less of the time than they did before! (Now Toshiba picks HD DVD $10/11$ of the time as opposed to $4/5$ of the time initially.)

This sounds very discouraging to Sony. They desperately want to adopt Blu-ray—more than Toshiba wants to go with HD DVD—and yet this revelation causes Toshiba to favor its preferred choice even more. But Sony's change in utility, as well as Toshiba's deviation from their original strategy, will still alter Sony's expected payoff. Let's compute it: Sony still mixes ($4/5$, $1/5$) while Toshiba's mix is ($1/11$, $10/11$). Sony's expected utility is as follows:

$$(4/5) * (1/11) * 10 + (1/5) * (1/11) * 0 + (4/5) * (10/11) * 0 + (1/5) * (10/11) * 1$$

$$= (40/55) + (10/55) = 50/55.$$

Happily for Sony, even though Toshiba avoids Blu-ray even more than they did before, Sony's average utility of $50/55$ does show an increase from the original value of $4/5$. And, by the way, you can check that Toshiba's average payoff will remain as it was.

Before wrapping this example up, consider one more important lesson: it would seem that the only way the players would both have access to the same information is to have shared it. But now that you are aware that changing the payoff values will affect the mixed strategies each company employs, the values that the players reveal to one another could be subject to strategic manipulation.

Let's now probe a little deeper into the equilibrium concept and try to sort out "good" equilibria from "bad" ones.

Narrowing Down the Solution Concept

The Nobel laureate Reinhard Selten had an inspiration when he considered the problem of multiple equilibria. Many (but not all) nonzero sum games have more than one equilibrium to choose among, and as you have seen, some of them are better or worse for one or both of the players. On top of that, certain equilibria seem to make more sense than others. Selten's idea was to try to refine the equilibrium notion in order to keep the sensible solutions and discard the doubtful ones.

To provide a foundation for his idea, let's first discuss the idea of credibility. If a player communicates a certain intention, are they to be trusted? A good vehicle for introducing the role of credible statements is an extensive form game called the Centipede game.

The Centipede Game

Here's how the Centipede game works: there are two players and an initial pot of money. Each player has two strategies available: take or pass. Player 1 can choose to take a certain proportion of the initial pot, leaving the rest for Player 2. Alternatively, Player 1 can pass. If she passes, the pot doubles.

Now it is Player 2's turn. He can take that same percentage of the bigger pot, leaving the rest for Player 1, or he can pass, in which case the pot doubles again. The players alternate, and whenever a player decides to take, the two shares are given out and the game ends. The game is played at most a certain predetermined number of rounds. If all plays are pass, then the amount in the pot after the last play is distributed according to the proportions favoring the next player. Two rounds of this game are illustrated in the following figure.

NO BLUFFING

The Centipede game can have 100 (or even more) plays. In that case, you can imagine the diagram representing the game having 100 "legs." Hence the name!

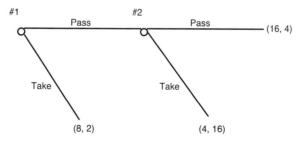

The Centipede game.

In this game the initial size of the pot is $10, and the proportion of the pot a player gets when playing take is 80 percent.

Suppose that before play began, Player 2 said to Player 1, "if you pass on your move, I'll pass on my move and you'll get $16 instead of $8." Player 1 might say in response, "what's in it for you?" Player 2 continues, "I'll get $4 instead of the $2 you would have left me on the first play."

Now suppose Player 1 frowns quizzically but decides to trust Player 2. She plays pass and the pot doubles to $20. What's to stop Player 2 from reneging on his promise and taking $16 on his move?

The trouble with Player 2's promise is that it is not rational. And since passing on his turn is not in Player 2's best interest, Player 1 will realize that the promise was not really believable to begin with. (As an aside, the issue of how people actually play this game is itself interesting.)

At the last play of the game, since the best strategy for Player 2 is to take, his opening promise to play pass was not believable. This, in turn, will affect how Player 1 will play.

The point of this Centipede game analysis is to begin to identify which strategies are plausible and which ones are not. Player 2 might offer to pass on his turn but Player 1 discovers that such a promise is not believable. This, in turn, allows Player 1 to prepare for the fact that Player 2, if given the opportunity, would most likely play take.

Subgame Perfection

Continuing with this credibility theme, Selten illustrated his idea in part with a version of the Market Entry game. The idea is this: imagine a Goliath of a company—for example, Walmart or Microsoft—and a start-up company, which we'll call it Upstart. In the game, Upstart will try to enter, one by one, different markets dominated by the Goliath.

In the first market (whether a product market or a geographical market), Upstart can choose to enter or to stay out. If Upstart enters, Goliath can play hardball (advertise heavily, engage in a price war, and so on) to fight Upstart or it can accommodate and let Upstart share some of the market.

The second round is more of the same. Regardless of what was done in the first round, once again Upstart can enter or not, while Goliath can fight or accommodate. The game can carry on for a number of rounds. The following extensive form diagram more precisely defines the game.

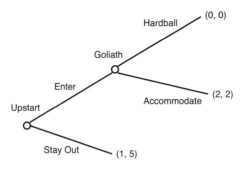

An extensive form diagram of the Market Entry game.

The branching that takes place to represent Goliath's strategies is a part of the larger game tree. Game theorists call the game originating at such a node in the game tree a *subgame*.

DEFINITION

A **subgame** is simply a well-defined game within a game.

Each player gets one move. Upstart goes first. In considering the strategy to enter the marketplace, Upstart would be concerned about Goliath playing hardball. But playing hardball is not rational, since Goliath gains more from accommodating Upstart. Therefore, working backward, Upstart sees that entering will lead to a payoff of 2, while staying out yields a payoff of just 1.

This line of reasoning means that the only equilibrium in the game that makes sense is (enter, accommodate), leading to the payoff (2, 2).

Now let's see what the normal (matrix) form of the game looks like.

		Goliath's Choices	
		Hardball	Accommodate
Upstart's Choices	Enter	(0, 0)	(2, 2)
	Stay Out	(1, 5)	(1, 5)

Normal form (matrix) diagram of the Market Entry game.

In this matrix, the entries in the bottom row are both (1, 5) because if Upstart stays out, it doesn't matter what Goliath had intended.

This representation of the game has two pure strategy equilibria: (enter, accommodate), and (stay out, hardball)! The second equilibrium amounts to Goliath being committed to fighting Upstart, but Upstart backing out. But as we have already seen, this equilibrium makes little sense.

Essentially, we subject each of the different equilibria in this game to a test: Is the equilibrium credible? In other words, are the strategies that lead to a particular equilibrium outcome rational? You have just observed that (stay out, hardball) involves a threat that is harmless, since it is against Goliath's best interest to fight Upstart. Therefore (stay out, hardball) has no credibility.

You can now eliminate this equilibrium, which leaves you with the more plausible equilibrium (enter, accommodate). The equilibrium (enter, accommodate) is credible because once Upstart enters, Goliath's best strategy is to accommodate. Selten called equilibria like (enter, accommodate) that satisfy our credibility test *subgame perfect*.

DEFINITION

A **subgame perfect equilibrium** is a set of equilibrium strategies that provide a Nash equilibrium in every subgame of the game.

The way to find these more robust equilibria is to work backward the way we just did with Upstart versus Goliath. I used the word "robust" because these equilibria hold up to scrutiny: the strategies are rational in every subgame.

At this point you have the basics of nonzero sum games under your belt. The next chapter considers a variety of nonzero sum models to help you appreciate the insights they offer.

The Least You Need to Know

- Nonzero payoff sums vary across outcomes.
- Nash equilibria are best reply strategies.
- Every nonzero sum game has a Nash equilibrium.
- Subgame perfection keeps the credible equilibria and tosses the rest.

More Nonzero Sum Games

Chapter 5

In This Chapter

- Variations on the Prisoner's Dilemma
- Understanding the classics
- Making sense of the solutions
- Softening mistrust

In Chapter 4 you learned about one of the classics of game theory—Nash equilibria for nonzero sum games. This chapter introduces you to more classics of game theory. These games are classics for a reason: they represent fundamental human conflicts in which self-interest can be ruinous.

Beyond just absorbing how these traditional game models work, you will also begin to see how the standard analysis yields some rich insight into how to behave—and get others to behave—in certain situations.

Conflict and Cooperation

The Coordination game you learned about in Chapter 4 serves as a good introduction to the mixed motives inherent in so many human interactions. The primary objective of the problem is to coordinate your actions with another's. This in itself is not insurmountable. However, the secondary aim, to achieve the best outcome, creates a predicament.

In the games that follow, both players are still attempting to reach their most favored outcome, but these games feature some insidious turns of events. Each game illustrates the mixed motives of the players—driven by incentives to cooperate, capitulate, or compete—and how game theory enables you to recognize the tension among those motives.

The Prisoner's Dilemma

Even if you have never personally been involved in the criminal justice system, I suspect you may have watched some television dramas in which a serious crime has been committed, and a couple of bad guys are caught and interrogated separately by law enforcement officers. The usual drill is that the police do not have enough evidence to convict one (or both) of the suspects, but in the separate interrogations they try to get the suspects to turn on each other.

If the suspects both keep mum, the police will only be able to convict them of a minor offense with relatively little prison time. But the interrogators are often able to get the individual suspects to plead to the crime by persuading each one that his accomplice has ratted on him. When the suspects rat on each other, they each get major prison time, but if one rats and the other stays mum, the rat ends up doing very little time and the other suspect takes the full rap for the crime.

Each player can either hold out (not confess), or rat on (implicate) the other one. You will see shortly how these two strategies have analogues in other situations. But for now, I will present the suspects' strategies as well as their payoffs in hypothetical prison time.

> **NO BLUFFING**
>
> The game known as the Prisoner's Dilemma surfaced in a paper written in 1950 by mathematicians Merrill Flood and Melvin Dresher of the RAND Corporation. The original game had nothing to do with suspected criminals. The catchy story was devised shortly thereafter by Albert Tucker, an important Princeton University mathematician and game theorist.

	Suspect 1	Suspect 2 Hold Out	Rat
	Hold Out	1 Year, 1 Year	10 Years, 6 Mo.
	Rat	6 Mo., 10 Years	8 Years, 8 Years

Payoffs in prison time.

Before proceeding with any analysis, I converted the prison sentences into numbers on a scale. Don't take the actual prison sentences too literally, however; it doesn't matter whether you use exactly 10 years as the sentence when one suspect takes the full rap for the crime, or whether the rat gets 6 months, 8 months, or just 4.

What does matter is the relative comparison of the prison terms. The worst individual payoff is when just one suspect is implicated; the second worst payoff is when both rat on each other; better is when they both hold out; and the best individual payoff is when one rats and the other one then takes all the blame.

Since the Prisoner's Dilemma, or PD for short, is not really about crime and punishment, let's convert the storyline into something wider-reaching and more useful. Instead of one strategy being termed "holding out," let's describe it as "cooperating" (with the other player). The other strategy, instead of being described as ratting on your friend, will be called "defecting." Finally, as is common in the vast literature on the Prisoner's Dilemma, let's use numerical outcomes on a scale of zero to five as follows.

		Player 2	
Player 1		Cooperate	Defect
	Cooperate	(3, 3)	(0, 5)
	Defect	(5, 0)	(1, 1)

The Prisoner's Dilemma using numerical outcomes on a scale of zero to five.

The payoff of zero is ingloriously called the "sucker's payoff" in the literature, while the payoff of five is the "temptation."

Since the actual prison terms have some leeway, what properties of the numerical payoffs make a game a Prisoner's Dilemma? Notice the ordering of the payoffs from best to worst: 5 > 3 > 1 > 0. The first condition is that the numbers in the matrix must obey the ordering just described. Let's check that they are consistent: the best payoff is the temptation (= 5), the next best is mutual cooperation (= 3), followed by mutual defection (= 1), with the sucker's payoff (= 0) being worst.

There is another requirement, which has to do with the results when the game is played repeatedly. Right now we are only considering the PD to be a one-and-done game, but later on we will delve into repeating it over time. In that case, to be a Prisoner's Dilemma, two times the mutually cooperative payoff must be at least as good as alternating between the temptation and sucker's payoffs, which in turn must be at least as good as doubling the mutual defection payoffs.

The mathematical condition for this requirement is as follows: $2 * 3 > (5 + 0) > 2 * 1$.

You want to make sure, for example, that repeated mutual cooperation is better than alternating between cheating the other player and then being cheated. Otherwise we would have the perverse situation in which the players' idea of cooperation was to take turns cheating each other!

So finally, here's the dilemma: Looking at the payoff matrix, you can see that defecting is a dominant strategy for both players. This means there is a Nash equilibrium in pure strategies with the outcome (1, 1). Of course, (1, 1) is a lousy payoff, especially considering that if both players were to cooperate with one another, they would achieve (3, 3).

Unlike the Coordination game, however, there is only a single pure strategy equilibrium. The more favored outcome (3, 3) is unfortunately not an equilibrium because either player (unilaterally) can improve by switching their strategy from Cooperate to Defect, obtaining a payoff of 5 instead of 3.

It's possible to interpret the strategies as being "nice" (cooperating) and "not nice" (defecting). But be careful not to confuse the players' rationality—doing the best for themselves—with higher-order goals like "doing the right thing."

> **BET ON IT**
>
> In nonzero sum games, the players can't be concerned with the payoffs the others get. For example, in the Prisoner's Dilemma, you are thrilled when moving from a payoff of three to five and you don't give a hoot whether the other player does worse or not.

The dilemma, then, is not a moral one, so don't ascribe any other utility to the strategy pairings except for the numerical outcomes in the table. Clearly, Defect is a dominant strategy for both players, and the outcome (1, 1) is not only a Nash equilibrium but the only Nash equilibrium. However, given that when both players cooperate, they are better off, we need to ask whether there exists a way to implement the mutually cooperative outcome.

One idea that might occur to you is that both players could discuss the matter beforehand, if that is possible, and agree to cooperate. While this sounds promising when dealing with honorable parties, such an agreement suffers from the same credibility problem observed in the Centipede game (see Chapter 4). Defection is a dominant strategy, period.

So given the assumption that your opponent is rational, you simply cannot expect him to cooperate, and even if he did, your best strategy is to defect anyway. Now let's have a look at some real-life examples of the Prisoner's Dilemma and see whether there is some practical way out.

Joint Ventures

Consider two executives. Executive A is in charge of a product that has great potential. But A's company is missing a critical element needed to bring the product to the marketplace. Perhaps they lack the ability to distribute this particular product, or they lack a complementary technology that is necessary for success.

This is where Company B comes in. Firm B has the competencies that A needs. Executive B has been in contact with Executive A and they both agree that engaging in a business venture that would bring together their complementary inputs would surely succeed. While this sounds marvelous to both parties, they weren't born yesterday. Each executive recognizes that a close collaboration with the other company presents some risks. The other company might steal trade secrets. Or they might simply discover enough about their partner's operations or technology to be able to compete in that sector in the future.

Is this business situation a Prisoner's Dilemma? The real question is, do real-life executives perceive the situation as such? It turns out they do. Each company in the hypothetical joint venture has two basic strategies: to fully cooperate with the partner company or to defect. Defection in this case would mean that while they cooperated on the surface, they cheated on their partner (stole industry secrets, etc.) as well.

If the perceived payoffs to real-world executives who have engaged in joint ventures can be shown to match the PD pattern, then it's safe to assume that the executives would strategize in the same way, even if they never actually heard of the "Prisoner's Dilemma" game.

It turns out that a survey of executives in the 1990s confirmed that they indeed view joint ventures with the same suspicions that we developed for the Prisoner's Dilemma. The executives, when asked to compare the various outcomes for the types of strategic actions, all gave responses that matched the PD pattern: the best outcome for them was defecting while the other firm cooperated; the next best outcome was from mutual cooperation, and so on, down the line.

But surely, going into a joint venture, or "strategic alliance," with significant trepidation can't be a healthy thing. For that matter, how many potentially rewarding business alliances have evaporated on account of fear of getting cheated by the partner firm? In the final section of this chapter I suggest some remedies that might help executives carry out their plans with more success.

Advertising Wars

Here's another version of a real-world Prisoner's Dilemma. In 1970, tobacco companies and the United States government reached an agreement stating that beginning January 1, 1971, the companies would discontinue advertising cigarettes on television, and in return they were granted immunity in any lawsuits prosecuted in U.S. federal courts.

Think about the competition among the tobacco companies prior to this agreement. Consider two of the larger companies, for example, R. J. Reynolds and Philip Morris. Both companies fight tenaciously for market share. An obvious way to compete is to spend money on television advertising. If Philip Morris were to advertise while R. J. Reynolds didn't, the accepted wisdom is that Philip Morris would wrest some market share from R. J. Reynolds, and vice versa.

But what happens when they both advertise? Their ads would basically cancel each other out. Market share, and more importantly, sales, would not increase, and therefore, the companies' profits would be diminished by having to spend excessively on their ad campaigns. As you can see, the advertising game is a form of an arms race where both parties would be better off agreeing not to spend!

The essential character of the interaction is captured in the following game matrix. The entries are approximate annual profits for the different scenarios (1970 figures, adjusted for inflation).

R. J. Reynolds	Philip Morris Don't Advertise	Philip Morris Advertise
Don't Advertise	(280, 280)	(112, 336)
Advertise	(336, 112)	(151, 151)

The Cigarette Advertising Face-Off. Entries are profits in $ (millions).
(Adapted from Roy Gardner, *Games for Business and Economics*.)

As you can see in this matrix, the substantial amounts of money spent on advertising have huge impacts on the firms' profits. The resulting situation is indeed a Prisoner's Dilemma. Advertising (the equivalent of not cooperating) is a dominant strategy for both players, and they are left with the mutually unsatisfactory payoff of $151 million each.

The irony of this tale is that when the U.S. government put an end to cigarette advertising on TV, they did the tobacco companies a favor. The firms were forced into playing their cooperative strategies, and the bottom line was that all of them came out significantly better than before.

BET ON IT

In the cigarette advertising war, the players benefited from being forced into their cooperative strategies. This gives us insight into how to get out of the dilemma that PDs pose: a binding legal agreement might be just what the doctor ordered!

Resource Management

Did you ever go out to dinner with a bunch of friends, and end up splitting the bill evenly? Now consider such an evening out, except that you have an objective that you truly wish to do justice to: you're trying to maintain your diet.

So what's likely to happen at dinner? You hardly touch the appetizers that you're all sharing, you order a healthy main dish, and then you skip dessert. Congratulations on your admirable restraint. Your friends, however, all gorge themselves and you end up paying for their gluttony. When the meal's over, you sort of wish you'd have pigged out also—at least you would have gotten your money's worth.

Suppose, prior to the meal, you had attempted to persuade your friends of the merits of dieting. Good luck with that! It might have been noble of you to try to convert them to your austerity, but you cannot realistically expect anyone to start dieting during an evening out.

Before we even talk Prisoner's Dilemma, let me confess that I cooked up the restaurant situation just so we could think about a geopolitical issue: carbon emissions regulation and reduction. The affluent Western countries might be willing to go on a diet, as it were, but what about developing countries? Emission reduction is a costlier pill for them to swallow.

Does the emission reduction situation fit the Prisoner's Dilemma pattern? It is not an easy task to answer this question, since we would have to decide what the relevant costs and benefits are over a certain time horizon.

Certainly, in any one year, the investment required to switch over to sustainable forms of energy production would outweigh any immediate benefits. In other words, defection is the easier strategy when planning for the short term. But by considering the dire predictions down the road, the push for sustainability becomes a rational enterprise.

In the meantime, any nation that pursues unilateral reductions is still at the mercy of others that continue to gorge themselves. With all the nations in the same boat, it would appear that some might be tempted to defect on an emissions agreement. In doing so, the defector, while failing to make any sacrifices, would still benefit from the good graces of others. This geopolitical football might sound to you like a multiplayer Prisoner's Dilemma.

Any discussion of resource management and the PD needs to take into account how the different players' actions are repeated over a long period of time. Any individual nation, for example, can veer from one policy to another as its elected officials are voted in or out of power. In a sense, there is no one Prisoner's Dilemma here, but a sequence of games instead. We consider repeated games in Chapter 22.

The Game of Chicken

I assume that mentioning the game of Chicken conjures up an image of crazed teenage boys in a destructive form of drag racing: driving their cars toward each other to see who "chickens out" first, or perhaps driving the cars in parallel toward a cliff's edge.

The game theory version captures this behavior. It's not that we want to emulate foolish teenagers; it's that numerous situations of personal or political interaction have the feature that the side that does not back down from a confrontation emerges victorious.

One appealing version of the story comes from Ken Binmore. You must have maneuvered a car in a situation where two lanes of heavy traffic narrowed into one. You are slightly ahead of the other car but they try to push past you. If you back off and they forge ahead, they end up in front. If neither of you forges ahead, you end up first, seeing as you were slightly ahead to begin with. But if you both forge ahead, you will get into an accident. Suppose we model the strategies and payoffs as follows.

	Car 2 Back Off	Car 2 Forge Ahead
Car 1 Back Off	(1, 1)	(0, 2)
Car 1 Forge Ahead	(2, 0)	(-1, -1)

The game of Chicken.

NO BLUFFING

Human beings are not the only wildlife who play strategic games. The "Chicken" matrix is actually related to a "Hawk-Dove" game describing animal behavior.

Examine the entries in the Chicken matrix. It is not a PD because the worst outcome is the payoff from both parties forging ahead. Since that outcome is different in nature from the analogous PD payoff, there ends up being no dominant strategy in Chicken.

There are two pure strategy Nash equilibria here: when one player backs off and the other forges ahead. But, as we know from our driving experiences, a policy of always backing off might leave us feeling inadequate, while a policy of always forging ahead might get us into trouble. What does the mixed strategy Nash equilibrium look like?

I will save you the algebra here. It turns out that the mixed strategy Nash equilibrium is for each player to pursue the strategy mix ($\frac{1}{2}$, $\frac{1}{2}$). Let's find out whether this mixed strategy is helpful. If both players mix ($\frac{1}{2}$, $\frac{1}{2}$) then the average payoff to each of them is as follows:

$$\frac{1}{2} * \frac{1}{2} * 1 + \frac{1}{2} * \frac{1}{2} * 0 + \frac{1}{2} * \frac{1}{2} * 2 + \frac{1}{2} * \frac{1}{2} * (-1) = 0.5$$

An average payoff of 0.5 may not sound so great, but it is better than the 0 obtained from the strategy of always backing off while the other player always forges ahead.

However, something doesn't appear to be quite right. If both players flipped a coin to decide whether to back off or forge ahead, then one fourth of the time they would end up in an accident! And since fender benders in these situations are not that common, something must be wrong with the model we've used.

Let's look at the payoffs: given that all the numbers range from -1 to 2, it makes no sense that the major hassle from getting into an accident is hardly any worse on our scale than the mild chagrin we experience when the other driver gets their way with us.

To make the model more realistic, we would have to adjust the payoffs to render them more representative of the utilities in real-life confrontations. Remember, Chicken is not really about cars edging ahead of one another. Below I present two matrices, each with an adjusted payoff in the mutually destructive entry. Let me redefine the strategies as "yield" or "don't yield."

(a)

	Yield	Don't Yield
Yield	(1, 1)	(0, 2)
Don't Yield	(2, 0)	(-10, -10)

(b)

	Yield	Don't Yield
Yield	(1, 1)	(0, 2)
Don't Yield	(2, 0)	(-100, -100)

The game of Chicken, with adjusted payoffs for the mutually destructive option.

In (a), the mutually destructive payoff is now -10, while in (b) the same payoff is much more catastrophic at -100. The mixed strategy Nash equilibrium for (a) is $(10/11, 1/11)$ for both players, while in situation (b) it's $(100/101, 1/101)$ for both.

Not surprisingly, as the mutually destructive outcome becomes more severe, the players will play less aggressively. In (a) the mixed strategy is to yield 91 percent of the time, while in (b), the mixed strategy is to yield 99 percent of the time. Doing the math shows that in each case, the average payoff under the mixed strategy Nash equilibrium gets close to 1, which is the payoff when both yield.

Finally, as with the Prisoner's Dilemma, the game of Chicken is also something you might contemplate as a repeated game. In fact, it is plausible that any single confrontation actually takes place in stages.

Let's view the multiple-stage approach with the automobile example. You can see that every time the cars both edge forward, they are playing the "don't yield" strategy. As long as the cars are relatively far apart, the other driver's edging forward is irritating, but there is little chance of an accident or other negative consequences. However, as the cars converge, the risk escalates and, at that point, according to the mixed strategy Nash equilibrium, the drivers better take their feet off the gas the great majority of the time.

Stag Hunt

The philosopher Jean-Jacques Rousseau, in describing "mutual undertakings" by primitive men, illustrated the tension between individual gain and group benefit by imagining the men hunting a deer. The hunters would have known that killing the deer would require everyone's participation. Yet, an individual hunter might be tempted to run after a passing rabbit, without regard to whether the joint effort was spoiled by his departure.

NO BLUFFING

Rousseau's eighteenth century contemporary, the Scottish philosopher David Hume, was not to be outdone when it came to parables on group benefit and individual action. His game involved two people rowing a boat. Both have to row in order to get somewhere, but one person might be tempted to ease up and try to get away without working much. This particular framing of Hume's parable is more in line with the Free Rider problem covered in Chapter 14.

To express the Stag Hunt game in matrix form, assume that there are just two players present. Each player can decide to cooperate (hunt the stag) or to defect (chase the rabbit). Then we have to decide what the payoffs are to the two players in the four different outcomes.

Thus far we have studied only symmetric nonzero sum games. In these cases, the payoffs to the players are mirror images; for example, in Chicken, the strategy pair (don't yield, yield) has a payoff of (2, 0) while (yield, don't yield) has a payoff of (0, 2). In the most general form of these games, the payoffs don't have to be symmetric. In the following Stag Hunt matrix the outcomes are shown as variables that are not necessarily symmetric. The row player's payoffs are denoted in caps, while the column player's payoffs are lowercase.

	Player 2 Hunt Stag	Player 2 Hunt Rabbit
Player 1 Hunt Stag	(S, s)	(C, r)
Player 1 Hunt Rabbit	(R, c)	(R, r)

The Stag Hunt game.

Let me explain the payoffs. When both players hunt stag they catch it. But even if they share the meat evenly, the benefits might be perceived differently. (One hunter might have a bigger family or value deer meat more.) Thus S may not equal s. When both decide to chase their respective rabbits, they each catch one but, once again, they might value rabbit differently so R may not equal r.

The equivalent of the sucker's payoff in the Prisoner's Dilemma would be the (asymmetric) payoffs C and c. This is when one player earnestly pursues the stag but gets cheated out of any benefit when the other player takes off after a rabbit.

How can the relationships among these variables model individual versus group tension? Payoffs S and s (when they catch the stag) have to be the best ones for players 1 and 2, respectively. Payoffs R and r have to be the next best, while getting cheated and left empty-handed is represented by the worst payoffs, C and c.

In formulating the matrix I wanted to emphasize two things: one, that payoffs in nonzero sum games need not be symmetric; and two, that for Stag Hunt, just like for Prisoner's Dilemma, the numerical entries can be tweaked within certain ranges and still express the essence of the conflict. Now let's use some values for Stag Hunt in the following matrix.

	Player 2 Hunt Stag	Player 2 Hunt Rabbit
Player 1		
Hunt Stag	(7, 6)	(-1, 2)
Hunt Rabbit	(2, 0)	(2, 2)

Asymmetric payoffs in the Stag Hunt game.

I have made the payoffs slightly asymmetric here. Both players value catching a rabbit equally (payoff of 2). But let's suppose that Player 1 has a bigger family, or has fewer nuts and berries stored up—this means that catching the stag is valued more highly by Player 1 than by Player 2 (as seen by comparing the payoff of 7 to 6); and that not catching anything, while valued at 0 for Player 2, is a slightly bigger bummer for Player 1 (who values this at -1).

What about the analysis? There are two pure strategy Nash equilibria: (stag, stag) and (rabbit, rabbit). Remember how in the Prisoner's Dilemma, pre-game communication was unlikely to help since the dominant strategy for both was to defect? Stag Hunt, however, is more amenable to pre-game communication and deliberation.

Stag Hunt is more benign than Prisoner's Dilemma because the cooperative outcome (stag, stag) is a Nash equilibrium that yields the absolute best payoffs for them both. You can imagine the players promising beforehand not to go chasing after rabbits, and actually keeping that promise.

Now, let's see some ways in which the perils of these classic games can be overcome. Here's an idea to get started down that path: what if you were facing a nasty Prisoner's Dilemma situation and, suddenly, a fairy godmother appears, waves a wand, and voilà, the interaction is now a kinder, gentler Stag Hunt!

Transform Lose-Lose Outcomes to Win-Win

Full disclosure: I don't believe in miracles. But I do believe that, sometimes, one game can be changed into another. Doing so is not alchemy or black magic. It is a matter of the players generating additional incentives—carrots or sticks—that serve to modify the payoffs in one or more of the outcomes. Adding conditions that make a particular outcome either more or less enticing can change the nature of the game and ultimately steer the players toward a mutually satisfying encounter.

BET ON IT

"Win-win" is an expression that has gained popularity, showing that our society has begun to achieve some understanding that personal and political interactions can transcend zero-sumness. It is truly enlightened thinking to grasp that one party coming out a winner does not require the other to be a loser.

Of the three games we have seen in this chapter, the Prisoner's Dilemma is the thorniest one. In Chicken, even when a player always backs off, the resulting outcome is not the worst one possible. Moreover, as we have seen, even when the mutually destructive outcome is catastrophic, both players can successfully employ a mixed strategy and arrive at an equilibrium that yields a significantly better outcome than when one player gives in to the other.

Stag Hunt, as we have just seen, is also less troublesome than Prisoner's Dilemma. Therefore, it makes sense to focus on the more prickly PD as we seek to nudge games in a favorable direction. To put it differently, the "lose-lose" outcomes in Chicken and Stag Hunt can be largely avoided in ways that I have already mentioned. Prisoner's Dilemma, though, is a tougher nut to crack.

Strong and Weak Prisoner's Dilemmas

In 1988, the mathematician and game theorist Anatol Rapoport offered the following insight into the nature of the Prisoner's Dilemma. Rapoport realized that people feel pressure to defect for two reasons: greed and fear. Greed spurs us to defect because of the desirable temptation payoff, while fear drives us to avoid being stuck with the sucker's payoff. If there is some way to reduce the temptation payoff in a real-life PD situation, and also a way to increase the sucker's payoff, then the urge to defect will be reduced.

Recall how there is a range of payoffs that satisfy the PD structure. Rapoport goes on to compare two different PDs, which have very different degrees of fear and greed. Note how greed and fear are two sides of the same coin of mistrust. You want to find a way either to reduce the mistrust that both parties feel, or, if that is not possible, to work around it. I use C for cooperate and D for defect in the following tables.

(a) Strong PD

	C	D
C	(3, 3)	(0, 5)
D	(5, 0)	(1, 1)

(b) Weak PD

	C	D
C	(2, 2)	(0, 3)
D	(3, 0)	(0, 0)

Strong and weak Prisoner's Dilemmas.

The strong PD (a) is the one we've already seen. But check out the weak PD (b). Defection is still the dominant strategy, but it is now weakly dominant—meaning while one payoff under D is strictly better than the corresponding payoff under C, the other payoff under D is only at least as good as, not strictly better than, the corresponding ones under C.

Suppose you were in a weak PD situation and believed that the other player was going to defect. Of course, D is still the dominant strategy, so you are rational to have this belief. In this case, you lose nothing by cooperating (your payoff is 0 in either case). But here's the point: the other player will be thinking exactly the same thing! So we've just determined that in a weak PD a reasonable, if not exactly rational, pair of players has a chance to achieve the mutually cooperative payoff.

BET ON IT

Weak Prisoner's Dilemmas are still PDs. And defection is still a dominant strategy in these situations. However, in the real world, if the temptation payoff is diminished to the point at which it hardly pays, some players will indeed cooperate, thus finding a way out of the dilemma.

This line of reasoning provides a glimmer of hope that the despair we associate with the PD can be overcome. But the onus is still on me to persuade you that incentives can be devised that actually change the payoffs. What else can be done?

Hostages and Enforcement

The word "hostages" in the heading must have caught your attention. The idea here is more literal than you might suspect. To see how it works, consider a joint venture PD between two companies. One way to implement the taking of hostages is for each company to hold some of the other company's assets until the joint venture has been completed to everyone's satisfaction.

Another way to execute hostage-taking is to create a side contract that is tied to the joint venture. For example, recall companies A and B in the hypothetical joint venture I described earlier in this chapter. Firm A's product had great potential but needed B's complementary technology or expertise to succeed. Perhaps A worries that B will steal its technology and market a similar product in the near future. To make the joint venture work, they might enter into a contract in which B is required to purchase a certain amount of the resulting output.

What is meant by enforcement can vary. One sense of enforcement would be a contractual agreement that implements the hostage scenarios. Another aspect of enforcement is for both parties to determine in advance what actions or consequences would constitute cheating, and to jointly create a binding agreement as to how the cheating, when detected, would be punished.

Keep in mind that taking hostages or entering into binding punishment clauses alters the payoffs. The temptation will be reduced if punishment for defection can be legally put in motion. Similarly, the sucker's payoff will be ameliorated if the party that was taken advantage of is recompensed at the cheater's expense.

Altogether, the preceding prescriptions are not necessarily easy to implement; however, for players stuck in a Prisoner's Dilemma, they provide some welcome get-out-of-jail-free cards.

The Least You Need to Know

- The Prisoner's Dilemma, Chicken, and Stag Hunt games typify a number of important situations that pit self-interest against cooperation.
- In the Prisoner's Dilemma, the mutually inferior outcome is the only equilibrium, leaving the mutually beneficial outcome basically unattainable.
- The game of Chicken can be dealt with to some extent by employing a mixed strategy.
- Stag Hunt entices players with a modest reward if they leave the group, but still retains a cooperative equilibrium.
- Even intractable games like the Prisoner's Dilemma can be morphed into agreeable ones when the right carrots and sticks are applied.

Part 2: How Information Affects Games

If you're playing chess or tic-tac-toe, all the previous moves have been played in front of you and you know the exact configuration of the game at all times. Now compare this to most situations, say, in the business world. Managers need to make decisions, but they're in the dark as to what other firms are doing. While they might be able to estimate their own profits in various market outcomes, they probably can't estimate what the different scenarios are worth to other companies.

It turns out that the range of information available to the players makes a huge difference in how they strategize. Moreover, in some games there's an informational asymmetry in which one player's move is meant to communicate something to the other player. This is called signaling.

Two examples: Is a used car at a dealership really worth the sticker price? Does your advanced degree really predict that you'll be a productive employee? After exploring the impact of the different types of information available, I examine the consequences of signaling in both its basic and more nuanced forms.

Ignorance Is Not Bliss

Chapter 6

In This Chapter

- Revisiting imperfect information
- Breaking through the fog of incomplete information
- Dealing with private information
- Incorporating states of nature

Sometimes in real life it's convenient to deliberately avoid obtaining certain information before you make a decision. But for most decisions, including those in game theory, it is truly a handicap not to be as fully informed as possible. This chapter explores the many ways information can be limited and what such limitations mean when playing games.

Perfect and Imperfect Information

I pointed out in Chapter 1 that everyday games like chess and tic-tac-toe are different from the strategic games that are the subject of this book. Recall the ordinary sense of how chess and tic-tac-toe are different from the strategic games like the zero sum game Market Share or Stag Hunt: when it is someone's "turn," or move, in the everyday games, all of the other moves up to that point in time are known to everyone playing. In other words, such games are played with all of the moves, and the resulting configurations, known at all times.

What about blindfolded chess players? The game is harder to play because instead of being able to see the board, they have to remember where all the pieces are. However, the idea is the same: none of the moves are hidden; the players are completely informed.

That's the ordinary sense of why everyday games are games of perfect information. The more formal sense uses the notion of an information set I introduced in Chapter 1. In games of perfect information, each information set in the extensive form diagram is a single node. There is no ambiguity about what has occurred.

In games of *imperfect information*, however, at least one of the information sets has two or more nodes. For example, in Two Finger Morra, the extensive form representation we looked at had two gray nodes. These were lumped together, and therefore indistinguishable, because they represent a player not knowing whether the opponent had put out one or two fingers.

DEFINITION

A game with **imperfect information** is a game in which at least one information set has two or more nodes. In other words, at least one of the players does not know the state of the game—which moves have been made—at all times.

By contrast, imagine being in the middle of a chess game. Suddenly you have to close your eyes. Your opponent moves and all you are told is that he or she either played knight to e5 or rook to c2. Chess is hard enough when you have access to all the information; not knowing what move your opponent played would take it to a new level.

Making the distinction between perfect and imperfect information may seem like an unnecessary technicality. But there is a good reason to make this distinction. Even though games like chess and Go are very interesting and difficult games for us mere mortals, these games lack excitement for game theorists because Ernst Zermelo proved in 1912 (before what we call "game theory" even existed) that these games are "strictly determined," meaning that the outcome, in theory, is known in advance.

Another reason to distinguish between games of perfect information and imperfect information is that the truly interesting games involving human interaction are games of imperfect information. If foreign policy, for example, were as easy as tic-tac-toe, by now we would have either achieved world peace or we would have annihilated each other.

POISONED PAWN

Just because perfect information games have a known outcome in theory doesn't mean they're easy to solve. In games like chess and Othello the number of possible moves is astronomical, and dissecting the entire game tree is way beyond the reach of even the fastest computers.

So, to summarize, with perfect information each player knows every move the other has made. With imperfect information, at least one of the players is in the dark as to what the other player has done or is doing at that moment. In Rock-Paper-Scissors you have to throw rock or paper or scissors without knowing your opponent's simultaneous move. In Chicken, you have to act aggressively or back off without knowing the strategy the other player will take.

I hope that this distinction is clear. But now let me provoke you with a new kind of uncertainty: suppose you are involved in a real-life Prisoner's Dilemma (PD) but you don't even know what the outcomes really mean to the other player. Does the temptation tempt them as much as it tempts you? How afraid are they of the sucker's payoff?

The Fog of Incomplete Information

With imperfect information you know your' opponent's set of possible strategies and the payoffs, but you don't know which strategy they are using. But here's a new twist: suppose that now you do not know what the payoffs are to the other player. In a very real sense, this new uncertainty leaves you baffled as to exactly what game you are playing. Games with this feature are called games of *incomplete information.*

DEFINITION

A game of **incomplete information** is a game in which at least one player does not know the payoffs to the other players.

As an example of a game with incomplete information, let's continue with the idea that you are involved in a Prisoner's Dilemma (see Chapter 5 for the PD payoff matrix). You know exactly how to evaluate the four different outcomes for yourself. However, you don't know just how badly the other player assesses the sucker's payoff. Your rating of the sucker's payoff for yourself is 0. You figure that the opponent might rate his or her sucker's payoff as 0 or 1.

Because you are uncertain as to what this outcome is for the other player, you can't immediately apply the type of solution analysis you learned for games of complete information. The fog of incomplete information is illustrated in the following two tables. You are the row player.

	(a)	C	D		(b)	C	D
	C	(3, 3)	(0, 5)		C	(3, 3)	(0, 5)
	D	(5, 0)	(1, 1)		D	(5, 1)	(1, 1)

The Prisoner's Dilemma with incomplete information.

Game (a) above is the usual PD formulation. Defection is a dominant strategy for both players and therefore you expect to end up at the outcome (1, 1). But in (b), defection is only weakly dominant for your opponent. That is, if you defect, the column player is indifferent between cooperating and defecting.

Since you don't know one of the payoffs to your opponent, put yourself in the opponent's shoes. How do they know what the payoffs are to you? Your opponent might have a wildly different perception about several of the payoffs. While you recognize the game as a Prisoner's Dilemma, your opponent might have strikingly different evaluations of your payoffs. This in turn means that the interaction from his or her point of view might have an entirely different character.

One way to view the notion that you don't exactly know what the other's payoffs are is to say that you don't know what type of player you are up against. In games of incomplete information, game theorists formalize this idea by allowing each player in the game to be one of a number of different types, meaning that each type could have a different set of payoffs in the different outcomes. Each player knows his or her own type, but at least one of the players does not know the type of the others.

This example is reminiscent of the discussion in Chapter 5 of the difference between strong and weak PDs. In the present case, the fact that the column player in game (b) might not be so prone to defection makes no difference to you; defection is still your dominant strategy. But not even knowing for sure which matrix you're in could matter big time in a different situation.

When Information Is Asymmetric

Just now I raised the idea that both players might be in a fog regarding what the other's payoffs are and, even worse, the nature of the interaction in general. But what if it's just one player who is in the dark? Many games of incomplete information in the real world are one-sided.

Think about wandering down any tourist-traveled street in a major city—for example, Broadway in New York. You might see some dubious people hawking "Rolex" watches. We are all aware that these timepieces, which can be purchased for impossibly low prices, are counterfeits.

But there are numerous other counterfeit or faulty products that less dubious people try to pass off as genuine or in good condition. I only need to type "used car salesman" to get my point across. There are plenty of situations in which customers are at a salesperson's mercy. Such one-sided situations can be formalized as games of *asymmetric information:* these are simply games of incomplete information in which one player has complete information but not the other.

BET ON IT

We often engage in transactions with salespeople who know much more about the product they're selling than we do. Such asymmetric information situations include listening to stockbrokers who peddle doubtful stocks; purchasing pharmaceutical products on the Internet; and, of course, shopping the used car lot.

Fool's Gold

Let's delve into an example of a game (adapted from Roy Gardner) with asymmetric information. In Fool's Gold, let's imagine a merchant of precious metals who occasionally pawns off fakes along with the real goods. Every week, this hawker attempts to sell some jewelry he has kept from various estate purchases. Depending on what he randomly selects from his inventory, the handiwork is either genuine 14K gold with probability p, or is fake, with probability $(1 - p)$.

The fake pieces need some fixing up—for example, covering them with relatively inexpensive gold plating. Rigging the fake jewelry costs the merchant a modest amount of money, but he does a good enough job so that no one else can tell the difference. The genuine pieces can be sold as is.

Let's assign some specific figures to the various outcomes. Let's say that the merchant will sell the jewelry for $125, whether it is genuine or not. Suppose that fixing up the fake pieces costs $30 (in labor and materials).

The buyer will be sufficiently enticed by the jewelry to value it at $175. This means that they would have paid any amount up to $175 to purchase it. But if the buyer ends up purchasing the fake piece, they will find out eventually and it will only have a value of $25 to them.

In the extensive form game shown here, the first move—the random provision of the jewelry—is made by nature, just as described in Chapter 2. The information the players have is asymmetric, because only the merchant knows whether the jewelry is real or fake. This distinction is also called "private information" by game theorists. After "nature's" move, the merchant decides to sell the item or not. If the item is for sale, then the buyer decides whether to purchase it.

To summarize the game:

- The merchant's item, provided randomly, is genuine with probability p or fake with probability $(1 - p)$.
- The merchant knows the quality of the item.
- The merchant decides to sell or not to sell.
- If the merchant declines to sell, the game ends.
- If the merchant decides to sell but the item is fake, the merchant fixes up the item at a cost. Otherwise the merchant sells the genuine item as is.
- If the merchant sells, the buyer purchases or doesn't purchase.

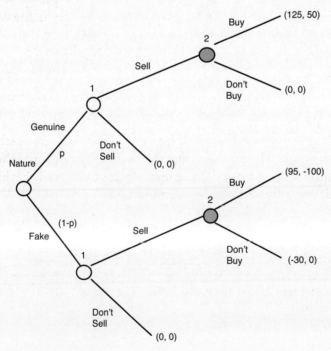

Fool's Gold in extensive form.

In this game, the merchant is Player 1 and the buyer is Player 2. These designations appear just above the nodes depicting their turn to play. Since the buyer cannot distinguish between the genuine and fake pieces, those two nodes are indistinguishable and therefore in gray.

The payoffs are easy to calculate. For example, when the buyer purchases the genuine item, her payoff of $50 is the valuation of $175 minus the price paid of $125. The merchant gets $125. If the buyer purchases the fake jewelry, it ends up being worth only $25. But after paying $125 for it, the buyer's net loss is $100. Here the merchant nets the revenue of $125 minus the $30 he spent to fix up the fake piece (hence a net gain of $95).

This game turns out to be pretty involved, so we won't solve it in full detail. But let's build up some intuition. Suppose the buyer, either from previous experience or from word of mouth, has reason to believe that the merchant sells mostly fake items. Since the buyer really gets burned when buying the fake items (a net loss of $100), without even doing the math we see that buying the item is surely too risky and the buyer will decline.

But what if the merchant almost always has genuine stuff, as far as the buyer has been able to discover? Suppose there is only a slight chance that the jewelry is fake. You can do the expected value if you'd like, but when the jewelry is almost surely genuine, the expected gain for the buyer will be positive.

What about the merchant? He is concerned with two things: his reputation, as measured by p, and his cost to fix up the fake item. If p is known to be low, then the buyer is not going to buy. In this instance the merchant won't go to the trouble of fixing up the piece if it is a fake one. But even then he's not selling anything and his profit is $0.

But even when p is moderately high, if fixing up the fake is very costly, then the merchant won't do it. The reason is that the buyer might balk and then the merchant would have spent a lot of money fixing up the piece to no avail. In this case the merchant will only offer the genuine items for sale.

BET ON IT

Fool's Gold, although a straightforward game, has a lot of complexity. Depending on the costs and probabilities involved, the equilibrium solution of the game can take on four different characteristics!

Additional factors influence how the game is played. The strategies that are pursued also depend on how the buyer values the item she ends up with. For example, how much is the genuine article really worth to her, and how devastating is it to get cheated? Let's summarize the different directions that Fool's Gold can take:

- **Market success.** This is a win-win situation. The merchant will only offer an item for sale if it is genuine. The buyer will buy because there is no fear of getting ripped off. This pure strategy equilibrium benefits both parties.

  An important observation about market success is that the merchants always do the right thing. The reason for this is simply that the economic conditions (the costs, probabilities, and so on) provide the right incentives to separate out the merchants into their "proper" roles. Merchants with good items use a different strategy from those stuck with bad items. This kind of equilibrium is called a separating equilibrium.

- **Market failure.** This is lose-lose! The merchant anticipates that the buyer will not buy. So regardless of whether the item is genuine, the merchant will not sell. And then there is nothing for the buyer to buy. Again, this is a pure strategy equilibrium, but without any benefits.

  Something else to notice about the market failure case: on account of the economics present, all merchants are lumped into the same boat. The ones with good items will not benefit from trying to sell, and neither will the ones who would have hoped to pawn off the fakes.

  So despite the fact that different types of merchants are present, they all use the same opting-out strategy and therefore cannot differentiate themselves in the market. This kind of equilibrium is called a pooling equilibrium.

- **Partial market success.** Here, the merchant always offers the item for sale, genuine or fake. The buyer will take her chances and buy whatever's on offer. On average, both parties benefit, but sometimes the buyer gets ripped off, and this lowers her average gain.

- **Slight market success.** In this situation both the merchant and buyer pursue a mixed strategy: for example, in this case the buyer sometimes takes a chance, and sometimes doesn't. This scenario is superior to market failure, but otherwise generates only small gains for the players.

Beyond Fool's Gold

There are lots of real-life examples of purchasing behavior in which we take our chances. Sometimes we don't even realize the chances we're taking—we believe we mitigate the risk when we rely on merchant reputation or brand names.

For example, suppose you purchase jewelry from a department store. They do not manufacture the jewelry and surely they don't fix it up à la Fool's Gold. But isn't it possible that the department store's buyers were themselves fooled by unscrupulous jewelry merchants?

Formal games like Fool's Gold are amenable to analysis that explains the different sorts of behavior that we see between buyers and sellers. But in some sense the true game being played does not end with the purchase. You must have wondered about people who proudly tote those fancy designer purses. Did they really pay $750 apiece for them? If not, they must be happy enough with an authentic-looking knock-off.

What about people who purchase prescription medication on the Internet? Every year, billions of dollars of counterfeit drugs are hawked around the world. In Fool's Gold, the analysis began with a random move by nature: delivering a genuine or a fake item. With prescription drugs, the situation is further complicated by the fact that some people are not cured by the real medication, while others get better despite the fact that they took essentially a placebo!

An interesting extension of the Fool's Gold model in the Internet prescription scenario would be to add a random move by nature at the end of the game, representing the efficacy of the drug after it was taken.

Another layer of complexity can be added by having the merchant offer different goods at different prices. The price itself would then act as a signal of the good's quality. The next chapter investigates such signaling games. To close out this chapter, though, let's consider the standard approach to solving a game of incomplete information.

Breaking Through the Fog

In Chapter 4 I introduced the Market Entry game, featuring an Upstart deciding whether to enter a market dominated by a Goliath. One way to represent the game is to use the normal form, which employs the same matrix used in Chapter 4.

		Goliath	
		Hardball	Accommodate
Upstart	Enter	(2, 3)	(4, 4)
	Stay Out	(3, 7)	(3, 7)

Upstart vs. Goliath represented in normal form.

Remember, the subgame perfect equilibrium strategy is for Upstart to enter and for Goliath to accommodate. The other (Nash) equilibrium is for Upstart to stay out while Goliath played hardball. This equilibrium, however, which might result from Goliath's threat to play hardball, is not credible. Upstart will realize that the only rational response to its entering the market is for Goliath to accommodate.

The analysis is sound, but now let's dig deeper and ask the following question: how does Upstart know what Goliath's payoffs are in the various situations? In particular, imagine the executive at Goliath who makes the decision whether to play hardball or to accommodate. Is it possible that Upstart has misjudged this individual?

Perhaps this executive gets a personal thrill out of playing tough. Or maybe Upstart has misjudged the economic incentives inherent in the different outcomes. Either way, the deeper analysis we now embark upon, based on the work of David Kreps and Robert Wilson, incorporates incomplete information.

> **BET ON IT**
>
> In games of incomplete information, at least one of the players is uncertain as to the other's payoffs. Since this makes the very game they're playing uncertain, the approach is to try to estimate how likely it is that they are engaged in one of the possible games.

The behavior modeled in the previous matrix pits Upstart against a weak Goliath. I am using the term "weak" because this Goliath will accommodate Upstart if the latter enters the market.

Now let's consider how a strong Goliath will counter a market entrant. Whether the payoffs will change due to a particular executive getting a thrill out of fighting, or whether the payoffs indicate something about anticipated future interactions with Upstart, the following matrix shows a different landscape.

Chapter 6: Ignorance Is Not Bliss

		Strong Goliath	
		Hardball	Accommodate
Upstart	Enter	(1, 4)	(4, 2)
	Stay Out	(3, 7)	(3, 5)

Upstart vs. strong Goliath.

Let's look at the payoffs. The big change comes if Upstart enters the market. The weak Goliath reaped a payoff of 4 when accommodating Upstart, but the strong Goliath only sees a payoff of 2 in this case. But by playing hardball, the strong Goliath—perhaps fighting now to discourage Upstart from challenging in other markets—obtains a higher payoff of 4.

The subgame perfect equilibrium solution with the weak Goliath was for Upstart to enter and Goliath to accommodate. The payoff was (4, 4). But with the strong Goliath, playing hardball is a dominant strategy and the only equilibrium is (stay out, hardball) with the dramatically different payoff of (3, 7).

Here's where the incomplete information comes in. Let's try to work out what strategy Upstart should take. If up against the weak Goliath, Upstart will enter. Against the strong Goliath, Upstart does best to stay out. A sensible thing to do is to ask how likely it is that Goliath is either weak or strong. Let's let p be the probability that Goliath is weak, with $(1 - p)$ being the probability it is strong.

Nobel Laureate John Harsanyi pioneered the following approach to solving such games of incomplete information. He suggested that we change the game of incomplete information into a game of imperfect information by using an initial move by nature. With probability p, nature gives us the weak Goliath, and with probability $(1 - p)$, nature selects the strong Goliath.

Now, Upstart does not know the actual probabilities here. But they can try to estimate p. In real life this would be accomplished through collecting data on Goliath's previous behavior, or trying to discern through other means how this adversary is thinking.

Let's let q be the Upstart's "guesstimate" as to the probability of being up against the weak Goliath, with $(1 - q)$ being the estimated probability of being up against the strong Goliath.

Here's the analysis: if Upstart enters, its average payoff is:

$$4 * q + 1 * (1 - q).$$

The first term in this expression represents when the weak Goliath accommodates, giving Upstart a payoff of 4. However, the strong Goliath will play hardball, which deals Upstart the payoff of 1 in the second term.

What if Upstart stays out? The payoff when staying out is 3 no matter what. So the question is, for what value of q is the average payoff from entering equal to the payoff from staying out? This is found when $4 * q + 1 * (1 - q) = 3$. The value of q that solves this is $q = 2/3$.

What this means for Upstart is if the probability of Goliath being weak is exactly $2/3$, Upstart is equally well off either entering or staying out.

But if the actual probability that Goliath is weak is less than $2/3$, Upstart is better off staying out.

What does this game theoretic analysis actually provide for us? It doesn't take an MBA to realize that Upstart wants to avoid a strong Goliath but would prey on a weak Goliath. But the value of the incomplete information model is that solving it gives us a threshold value to work with. In this case the threshold value is the probability of $2/3$ that Goliath is weak.

If there is a succession of upstarts, or a single upstart going up against Goliath in a succession of markets, we have a repeated game. Imagine that Upstart collects data on Goliath's strategies from game to game. Doing this will enable Upstart to revise the "guesstimate" value q. Since this kind of updating is typically done using Bayes' rule (see Chapter 2), such games are called *Bayesian games,* and the sort of equilibrium that we just worked out is called a *Bayesian-Nash equilibrium.*

> **DEFINITION**
>
> **Bayesian games** are games of incomplete information that one approaches by modeling uncertainty with a move by nature. A **Bayesian-Nash equilibrium** is a Nash equilibrium in a Bayesian game.

In summary, if the player who is in the dark can formulate a good guess as to what type the other player is, they can work out a satisfactory strategy.

The Least You Need to Know

- Incomplete information is worse than mere imperfect information. With incomplete information you don't even know exactly what the game is.
- Many buyer-seller games involve one-sided, or asymmetric, information.
- Depending on the economic conditions, games of asymmetric information can have drastically different solutions.
- You can solve games of incomplete information by using an initial move of nature to map out the different games.

Signaling and Imperfect Info

Chapter 7

In This Chapter

- Knowing when education is worth it
- Recognizing which signals are reliable
- Driving out the good with the bad
- Using signals as part of a general strategy

At a couple of junctures thus far, I have suggested that one or both players might profit from being able to send a signal of some sort. In this chapter I explain a variety of signaling games and help you decide whether signaling can lead to an advantageous outcome. In addition you get a chance to scrutinize and better understand certain commonplace behaviors by considering them as signaling strategies.

The Costs of Signaling

We all know how difficult it is to find the best person for a job. Often, dozens of applicants send in resumés that are chock-full of impressive credentials. The problem is, do these credentials actually indicate whether the candidates are indeed qualified? In the following analysis I set up a *signaling game* that goes some way toward answering this question.

To try to determine whether a job candidate is qualified, you would want to know what it took to achieve her various credentials in the first place. Some credentials might signal high productivity in the workplace, while other credentials might not.

> **DEFINITION**
>
> A **signaling game** is a game of asymmetric information where the first player sends a message to the second player. The first player knows her own type. The second player does not know the first player's type but will act after receiving the message.

For many years the entry-level ticket to a well-paying career has been a college degree. A degree might prove someone's dedication and intelligence, or at least demonstrate one's willingness to invest valuable resources toward the goal of self-improvement. Perhaps many degree programs provide skills that are valuable in the marketplace.

Here's how Nobel Laureate Michael Spence approached the problem of matching employers with job candidates. The employers want to hire productive workers while the candidates want well-paying jobs. Suppose you have a gut feeling that someone with a degree will be more productive than someone lacking a degree.

How much is a productive worker worth to an organization? Let's measure their output: suppose that a productive worker generates $100,000 of revenue for the company, while an unproductive worker generates less, say $50,000 worth.

Now consider a potential employee trying to figure out how to get their career started. Using the same dichotomy as the employers, let's assume everyone in the job market knows their own type: whether they are truly a productive or unproductive worker. This information is not known, of course, to the employers. What strategy should a particular individual take: obtain a degree prior to entering the job market, or else enter the job market straightaway?

We all know that getting undergraduate and advanced degrees requires a lot of time, effort, and money. We also know that at many institutions, degrees can be had for little more than paying the tuition and showing up for classes occasionally. So for now let's consider the problem of obtaining a degree as basically monetary. Students pay for their credentials both directly in tuition dollars and in foregone income.

> **BET ON IT**
>
> Getting a degree is one way to send a signal. But keep in mind that university degrees, like most signals, can't be obtained for free if they are going to be meaningful. There must be some kind of cost or sacrifice involved.

The employers are only interested in whether they hire productive workers. But since they can't attach electrodes to the heads of the job candidates to figure this out, they rely instead on resumés and credentials. Remember, in this example we're assuming that holding a degree is a signal of productivity in the workplace.

The question now is, how much does it cost to obtain a credential, or in other words, to invest in a signal that says "I'm productive" to the employer?

Spence assumed that it costs more for an unproductive person to obtain a credential than for a productive person. There are different ways to view this. One is that the naturally unproductive person requires more effort to reach the same goal. For example, the unproductive person ends up requiring eight years instead of four to get his diploma.

Without coming up with an exact number yet, let's suppose that a productive person needs to spend C to obtain her credentials, while it takes double that (equaling $2C$) for the unproductive person to get his.

What salaries will the employer pay? Suppose the employer believes that credentialed applicants will be productive, and that noncredentialed applicants will be unproductive. Going along with this, suppose the employer advertises a salary of $90,000 to applicants with a degree, and $45,000 to those without a degree.

Now we are in a position to figure out the following important things:

- Which applicants should seek a degree
- Whether the employers are actually hiring the right people at the right salaries
- How much the degrees themselves should cost

Separating the Men from the Boys

In the discussion of Fool's Gold in the previous chapter, you learned that Market Success was achievable under certain conditions. Under Market Success the merchants self-selected into those who sold their merchandise (when it was authentic) and those who did not offer it for sale (when it was fake). This self-sorting behavior was maintained by a separating equilibrium in that game. Can the same concept apply here in the job market? Let's do the math and see.

Do the Math

First, let's summarize the data in this job market:

- Productive workers generate $100,000 of revenue.
- Unproductive workers generate $50,000 of revenue.
- The firm pays $90,000 to applicants who have a degree.
- The firm pays $45,000 to applicants who do not hold a degree.
- It costs C for productive types to earn a degree.
- It costs $2C$ for unproductive types to earn a degree.
- The firm does not know the applicant's type.

Suppose a productive applicant is hired. (Remember, the employer does not know whether this applicant is productive.) Factoring in the cost of education, this person will net $90,000 – C$ if she has a degree, and $45,000 if she does not have a degree.

Similarly, an unproductive applicant will net $90,000 – 2C$ if he has a degree, and $45,000 if not.

It is cost-effective for a productive type to obtain a degree provided that $(90,000 – C) > 45,000$.

What about the unproductive type? This person is better off not getting a degree provided that $45,000 > (90,000 – 2C)$.

If we solve these two conditions together, we find that as long as the cost of education (C) is less than $45,000, the productive type is better off getting her degree. We also find that as long as C is greater than $22,500, it is not worth it for the unproductive type to obtain a degree.

The Analysis

The result is a remarkable separating equilibrium! As long as the cost of a college degree in the model is between $22,500 and $45,000, the two types of job applicants sort themselves perfectly. The productive types always obtain a degree, and conversely, the unproductive types never do. The signal that they both send is perfectly reliable. The employer will hire credentialed applicants at the high salary, and the uncredentialed ones at the low salary.

Finally, some interesting icing on the cake: this simple societal model also tells us how to price education in order to obtain a desirable outcome. Notice that a low price tag (say, $22,501) is more efficient for this society than a high price tag (say, $44,999).

> **POISONED PAWN**
>
> Don't make the mistake of taking any game theory model too literally. The separating equilibrium just described provides a compelling story, but you must resist the temptation to dash off a letter to the editors of *The Chronicle of Higher Education* telling them who should go to college and what prices to set for degree programs. Models are only as good as their assumptions, and before drawing conclusions for the real world you have to think hard about the legitimacy of making generalizations.

Pooling Everyone Together

While the separating equilibrium has some implications for society and how higher education could be designed, it is not necessarily representative of the role of education in the job market in the real world. For one thing, different degrees send different signals; their costs are different; some institutions have high standards while others do not; and so on.

As you might suspect, the perfectly sorted world we just fashioned is not the final word on this subject. It is possible that signaling will not work. Let's see under what conditions our ideal job market will break down.

First, it is reasonable to assume that some proportion p of the pool of job applicants is productive, while $(1 - p)$ of the applicant pool is unproductive. When we crank out the numbers, let's use a probability p of 0.40. (The general results are similar no matter what.)

Suppose an applicant has a degree. The employer assumes they're productive and offers a salary of $90,000.

However, when the employer is presented with an applicant without a degree, the employer assumes a toss-up: the applicant is either productive (with probability $p = 0.4$) or unproductive (with probability $[1 - p] = 0.6$).

Then the employer can figure out the average productivity of an uncredentialed applicant. This is (in dollars) equal to $100{,}000 * p + 50{,}000 * (1 - p)$, in general. Now let's do the math. It is similar to the previous example.

Do the Math

The average productivity of someone without a degree is $100{,}000 * p + 50{,}000 * (1 - p)$.

When $p = 0.4$, this is equal to $100{,}000 * (0.4) + 50{,}000 * (0.6)$, which equals 70,000.

So let's assume that the employer offers a salary to anyone without a degree equal to $65,000.

But a productive applicant who holds a degree generates $100,000 of revenue and will be paid $90,000. The cost of the degree is still C. Thus the credentialed applicant will net $90{,}000 - C$.

Let's put ourselves in the position of the productive applicant. Without a degree, they can make $65,000. When $65{,}000 > (90{,}000 - C)$, or equivalently, when $C > 25{,}000$, it is not worth it to obtain a degree.

The Analysis

The upshot of these calculations is that if it is somewhat expensive to obtain a degree, it is not worth it to the productive type. And remember that the unproductive type has to pay twice as much to get the degree. If getting the credential isn't worth it for the productive person, it certainly isn't worth it for the unproductive one.

So, in this case, when the cost of education is high (greater than $25,000 in the model), all applicants are lumped together in the same boat. They all select the same strategy, which is not to seek a degree. This is a pooling equilibrium, where the different types are indistinguishable to the employer. Since no one will obtain a degree, there is no signaling of quality and, consequently, the employer will offer everyone the low salary.

Signaling Quality

I just used the term *quality* as a more general way to connote "productivity" in the job market signaling game. Let's now depart from the job market and return to the previous intrigue of dubious merchants and used car salesmen.

Selling Lemons

Everyone knows that the moment a new car is driven off the lot it loses a big chunk of its value. Way before the car loses that new car smell, thousands of dollars have vanished, and it's not immediately clear why.

If all cars were perfect, the facts of car life would surely change. But not all cars are perfect: some are a source of everlasting worry and expense, and we call them lemons.

Nowadays the proportion of lemons is probably quite small. But the fact that they exist will persistently impact the market for used cars.

The market for lemons is another example of a game of asymmetric information. It is similar to Fool's Gold, except that in the used car market the vehicles can be offered for different prices. (Recall that in Fool's Gold all jewelry was offered for the same price regardless of authenticity.) The person selling a car knows its quality: they know the car is either reliable or a lemon. The buyer does not know the car's quality.

Before doing any analysis, let's assume that the cars in question are not very old. We need to ask why someone would sell his car in the first place. In particular, why sell a reliable car? There are all sorts of reasons: the seller has a new family situation and needs a different vehicle; he is moving to a pedestrian-friendly neighborhood; he absolutely has to have a new and higher-status car; and so on.

But we all know that once an owner realizes his car is a lemon, he will want to get rid of it: the cost and aggravation involved in its upkeep will be excessive.

The question now is, how to price the preowned car. Since lemons are somewhat few and far between, the expected value of used cars on the market should be pretty high—for example, at 80 percent of the new car price. This asking price provides a signal from the seller to the buyer.

BET ON IT

On the face of it, setting a price for a certain year and model used car would seem to provide a useful signal. Reliable cars would command a high price, lemons a low price. But what's to prevent lemon owners from jacking up their asking price? There's little or no cost to sprucing up the lemons so they perfectly resemble reliable cars, and therefore we do not obtain a nicely behaved separating equilibrium as in the education market.

To use some illustrative figures, suppose we're talking about a car that would cost $25,000 new. It is on the market used, and the asking price is $20,000 (80 percent of $25,000). If a buyer knew for sure that the car was perfect, she might indeed be willing to pay $20,000 for it.

But all buyers are naturally worried that they will get stuck with a lemon. Let's say that a lemon is worth half as much as a reliable car, or $10,000.

Suppose a buyer in our market knows that the great majority of cars are reliable. If they perform an expected value between a lemon (worth $10,000 with probability p) and a reliable car (worth $20,000 with probability $[1 - p]$) they obtain a number much closer to $20,000 than $10,000. Let's say the buyer makes a counteroffer of $18,000 for the car.

Who would accept such a counteroffer? Owners of reliable cars believe that their vehicles are worth $20,000. Therefore, relatively few such owners will take the $18,000. But owners of lemons would jump at the chance to unload their car on some poor unsuspecting person for $18,000. Because of this, the probability a car that sells for $18,000 is a lemon moves way beyond the original estimate p.

At the next stage of reasoning, no savvy buyer would offer $18,000 because they realize that there is an even higher chance of getting a lemon then they initially estimated. So maybe they think about lowering their counteroffer to $16,000.

Notice that this same logic applies to lower prices, too. At $16,000, even fewer owners of reliable cars would be willing to sell. Therefore, at $16,000, there is a still greater chance of getting a lemon. In response to this realization, the buyer would think about offering only $14,000, and so on. The result of this line of thinking is that at prices above $10,000, the reliable cars will not sell but the lemons would. Economists say that in such situations, "the bad drives out the good."

I realize that this argument is a bit convoluted, so let me recap the 10 steps of lemon logic:

1. Some proportion (p) of cars are lemons.
2. Lemons are indistinguishable to the buyer.
3. Owners set asking price equal to the value of a reliable car.
4. The possibility of getting a lemon sets the average value of a car below the reliable price.
5. Therefore the buyer's offer is lower than the reliable price.
6. Reliable owners would refuse the low offer.
7. The buyer anticipates such a refusal. The buyer's updated estimate of p is higher than previously thought.
8. To reflect the higher likelihood of a lemon, the buyer lowers their offer.

9. Now even fewer reliable owners will accept.
10. Go back to step 8 until the market collapses.

There is an elegant way to summarize this argument: For any positive probability p that a buyer would estimate, they would make an offer less than $20,000. Since no reliable owner would accept a price less than $20,000, the only cars that could be sold are lemons, and the only price anyone would pay is $10,000.

Another way to think of the market for lemons is similar to Fool's Gold but with two possible prices. Our world of used cars would be easier to navigate if there were a separating equilibrium like this: owners of lemons ask $10,000 and owners of reliable cars ask $20,000. But since there is little or no cost to fix up the lemons, why should the lemon owner signal low quality and end up with only $10,000 instead of $20,000?

BET ON IT

The usual analysis in the game theory, beginning with a famous paper by Nobel Laureate George Akerlof in 1970, is based on the sort of expected value mathematics just described. But there are other, subjective, reasons for making decisions. The fear of a lemon's aggravation might outweigh any economic analysis, even when the probability of getting stuck with a lemon is quite low.

Market Failure and Insurance Nightmares

It is possible that you just had a brilliant, cutting-the-Gordian-knot solution that can eliminate a lot of the uncertainty that plagues the lemons market: what about offering a warranty?

Great idea! What if you bought a lemon but your expenses would be covered, or you could simply return it and get your money back? Such a provision would certainly ease the tension in the used car market. Offering a warranty or something similar is a trustworthy signal that the product is high quality. For example, many reputable dealers invite people to see a CARFAX report, and by doing this the lemon problem seems to disappear.

Or does it? Let's briefly consider a completely different market, where it is not so easy to provide a reliable signal.

> **POISONED PAWN**
>
> Don't be fooled into thinking that warranties totally eradicate the lemon problem. A warranty often comes at a price: a markup on the cost of the item itself. For example, you pay a premium to buy the used car from a dealer as opposed to finding some vague seller on the Internet. But how trustworthy is the dealer? Caveat emptor!

Think about an elderly individual buying medical insurance. Suppose that for a person of a certain age, the average amount of money they will spend on medical treatment in one year is $10,000. Since insurance companies need to make money, the cost of a policy will have to exceed $10,000 or else their average payout will exceed what they take in.

Consider an elderly person in good physical condition. Their medical expenses on the average are relatively low, in other words, not likely to be anywhere near $10,000. Since the policy costs more than $10,000, these healthy folks will not buy insurance.

Now let's consider someone of the same age who knows he is in pretty bad shape. Outside of a few things that the insurance company could test for, this person's condition is going to remain as private information. Such people would expect to spend much more than $10,000 in a year (given that the average was $10,000) and a policy in the neighborhood of $10,000 would be a bargain for them.

Just like the market for lemons, the market to insure the elderly will break down. The average cost of the insurance policy is a good deal for those who are in bad shape (analogous to the auto lemons) but is a bad deal for those who are in good shape (analogous to the reliable cars).

Consequently, the only people who do purchase policies are those whose medical expenses are expected to be very high. The insurance companies would be ruined if they sold those policies. So the result is market failure: the cost to cover those in poor condition drives out the healthier folks, and in anticipation the insurance providers withdraw from the market.

No wonder that in the United States, for example, the government decided to provide a medical expense warranty of sorts in the form of Medicare, which was instituted in 1965.

Status Symbols as Strategies

There are numerous other "games" people play where sending signals is an integral part of the action. The sorts of "games" I'm thinking of are hard to capture with crisp and clean game theory models but they are nevertheless full of strategic interaction and deserve mention.

Expensive Signals

Do you wear designer clothing or accessories? If so, why did you purchase them? I doubt that your choice was made because designer merchandise is better made or more durable. We engage in such "branding" behavior all the time: we attend certain universities (and wear their sweatshirts proudly); we buy certain make cars; we spend three times the necessary amount of money to have a logo on a shirt; and so on.

Clearly, branding is successful because the identifications are meaningful to so many of us. The meaning surely involves signaling. Perhaps you manage your appearance in order to identify with a certain movement or sensibility, in the hope of attracting like-minded people. But what are you signaling with the collegiate sweatshirt, the BMW, or the Sub-Zero fridge?

You may retort that the designer gear simply makes you feel good about yourself. But why is that? There must be some underlying game, some sort of competition. The question is, what are people competing for when they wear designer brands?

This is not the place to further dissect such a complex issue. But surely the motivation to adopt status symbols is to project some kind of strength, power, or position. If you can blow $100 on a mere T-shirt, or spend an additional $2,000 on a refrigerator, isn't that a signal of vast monetary resources? And what does that imply, in turn?

There are plenty of other symbols of wealth, but one in particular comes with a whole subtext. Diamonds have become a rather universal symbol when given as gifts. The meaning in that context is often a signal of investment: "I am investing in this diamond to prove that I am willing and able to invest superior resources in my potential offspring." By the way, if that last sentence rings unfamiliar and obscure to you, bear in mind that some of the signaling we do is not conscious!

Equally interesting, our strategic signaling as individuals has become adopted by organizations in the business world. Perhaps the most celebrated such exploit was the "Intel Inside" campaign that was launched in 1990.

The Handicap Principle

In this book I soberly remind you, on occasion, that humans are still animals at heart. Already an entire generation of biologists has explored how the rest of the animal kingdom engages in behavioral games that are hardly different from much of what we get up to. The biological analysis is not only fascinating but also frequently offers insight into the behavior of *Homo sapiens*.

You must have watched at least a couple of TV programs that are beautifully photographed in remote parts of the rain forest where the creature diversity in appearance and behavior is utterly astonishing.

Certain animals and insects have evolved remarkable camouflage. The utility of these designs is obvious. These creatures like to hide, and their signaling seems like, well, the very antithesis of signaling. Other species have developed brilliant and attractive colors and appendages that would win signaling prizes, if they existed. But certain appendages appear to be so cumbersome and costly that scientists are not sure of their function.

Perhaps the most celebrated items in the cumbersome appendage category are peacock tails. Male peacocks drag around ridiculously long tails, replete with those marvelous feathers. The tails are absolutely a hindrance, and it is not clear how they could fit an evolutionary purpose.

In 1975 biologist Amotz Zahavi proposed that peacocks evolved longer and longer tails as a handicap that offers visual proof of the bird's biological fitness. Remember how I noted that for a signal to be worth anything, it must be costly? The idea with the peacocks is that, by proving that they are indeed able to drag those tails around, they thereby offer a signal of their superior fitness. The long tail would be too costly for a lesser bird to manage.

POISONED PAWN

Obviously, peacocks don't sit around wondering how they're going to grow extra-long tails. Strategic behavior need not be conscious! In fact, in the case of peacocks, the "behavior" of growing long tails isn't even a behavior, much less a strategy. The long tails are simply mutations that have been selected for over many generations. Nevertheless, viewing such mutations as strategies in the game theory context can be revealing.

This theory is controversial, but aside from being a startling and intriguing idea, over the years Zahavi's handicap principle has found its way into explanations of other behaviors, human as well as animal. The more general version is that anyone who can afford to waste vast resources must have extraordinary strength or resources to begin with. The handicap principle has also been shown to function as a game-theoretic equilibrium notion called an evolutionary stable strategy (see Chapter 17 for details).

The Least You Need to Know

- Signals generally have to cost something to be worthwhile.
- In certain markets, signaling is a reliable way of honestly differentiating among candidates of differing quality.
- Some markets break down to the point where the buyers cannot distinguish between the good and the bad.
- Offering warranties can help ferret out what's worth purchasing.
- The cost element of sending a signal extends to many strategic behaviors in human interaction and in the animal world.

Nuanced Messages

Chapter 8

In This Chapter

- Spending money as a signal
- Sending oblique signals
- Sending threatening signals
- Using volatility as a signal

In the previous chapter you looked at concrete signals such as a job applicant listing a college degree on a resumé or a merchant offering an item for sale at a certain price. In this chapter you continue to explore the role of signaling in strategic behavior. Some of the game situations here are less well-defined than before but still worthwhile to study.

Reading Between the Lines

There's an old expression, "it takes money to make money." One sense of this saying is simply that small investments don't generate much return while large investments do. But that's just the mechanics of investment returns. Doesn't money, or the lack of it, send a signal all on its own?

Signaling to Investors

Suppose you are considering a certain lawyer to represent you. When you visit her office, it is shabby and run-down. What kind of signal is that? You will be wondering, how good a lawyer is she if she can't even afford nice furnishings in her office?

Ultimately, you'll probably decide to seek counsel from the ritzier lawyer in the well-decorated suite instead.

To avoid this sort of rejection, the shabby lawyer would do well to spruce up her workplace, even if she has to borrow the money to do so. Once again, notice the value of sending a costly signal.

This principle would seem to be fairly prevalent. Suppose you are toying with the idea of investing some money in a certain company. You obtain its annual report. What would you look for?

I'm not thinking about the glossiness of the paper or the brilliance of the artwork in the annual report. What sort of data would you look for? Remember, you want to invest in a bright future, not in a bright past, so certain measures of past performance might be misleading.

It might occur to you to look at the dividends paid as an indicator of the future health of the firm. A dividend increase is promising investment news, the standard theory being that a company will only increase its dividend if the board believes that corporate earnings have permanently shifted upward.

But in fact, this signal—the dividend increase—is merely a strategy in a game of incomplete information. Maybe the firm is nothing more than a sophisticated Fool's Gold merchant, paying out higher dividends today in order to attract some more investment money tomorrow. (It takes money to make money.) In fact, some academic research shows that there is no straightforward relationship between dividend increases and future earnings growth.

There are, of course, many more corporate signals than just dividend changes. You could look at stock splits, charitable contributions, and even somewhat subjective information such as what's in the mission statement or the CEO's letter. But the essential message is that such communications, even of pure data, are subject to the same kind of scrutiny you gave to the dubious merchants in Chapter 6.

Developing a Reputation

In Chapter 6 I noted how the merchant in Fool's Gold might worry about his reputation. That worry had nothing to do with feeling remorseful—the meaning is simply economic, derived from the game model. Recall that p was the probability assigned to the given item's authenticity. If the buyer believes that p is sufficiently low, the buyer will not buy, and then the seller loses out, too.

As you learned, one way to adjust these probabilities, or at least the rough estimates, is to observe data from repeated activity. Each action taken is one more signal that gets added to the whole. Ultimately, the profile obtained from these observations is exactly what I mean by "reputation." If the merchant wants to attract future business, they'd better make sure that they behave most of the time, or else a low p will drag them into market failure territory.

So a reputation is simply the probability that a player is of a certain type or will play a certain strategy. Depending on the circumstances, we could be talking about a reputation that a player is honest, a reputation that they are cooperative, or that they would sell you a fake.

Costly Commitment

You have seen how signals need to be costly in order to be effective. Now let's apply the same idea but in a more nuanced setting.

Suppose you want to improve your reputation for trustworthiness. One way to build a good reputation is through good deeds. But what if you don't have months and years to repeatedly prove your good faith? One proxy for this sort of good repute is to demonstrate faith of a different kind: you might want to join—and become active in—a house of worship.

A religious association might signal, with more or less reliability, your reputation in more secular spheres. As with other signals, though, talk is cheap. You can't join a church, synagogue, or mosque one minute and try to leverage its good graces the next. You would still need to put some sincere time and effort in that institution.

Religious association or not, what if you were trying to raise money for a new product that you have just developed? The BBC television show *Dragon's Den* features novice entrepreneurs who try to raise funds from a panel of hardened businesspeople. In one riveting episode the panel grilled an inventor on how much of his own funds he was investing in his project. They were furious to discover that the entrepreneur was not willing to invest in his own scheme, while hoping that he could raise all the necessary capital from them.

Had the entrepreneur been willing to risk his own money or property in his project, the panel would have taken it as a sign of confidence in the outcome. Without that costly commitment, however, the panel rightly balked, seeing as the entrepreneur was willing to gamble only with other people's money.

BET ON IT

Commitment signals come in all shapes and sizes. Countless military commanders have burned ships or bridges. This says to their troops, "there's no going back." It also says to the enemy, "we are not only confident that we can beat you, but we will be fighting that much harder." So the quick lesson is that designing some sort of irreversibility into the game not only prevents withdrawal, but sends a strong signal to the other party.

A final example of costly commitment speaks to the relationship between employers and employees. One lamentable aspect of the current labor market is the dearth of loyalty. Employers are loathe to invest heavily in a new employee because that person might jump ship before the year is up; conversely, new employees feel equally vulnerable since they might be tossed out the door just as fast as they were wooed in.

One kind of costly commitment on the part of employers is to invest in the employee: to provide training programs, or to shorten the probationary period. These are signals that the employer is investing in the employee for the long term. On the other side of the coin, employees could be invited to design incentives that depend on the longevity of the employee's tenure at the organization—deferred bonuses, for example, that would reward an employee's loyalty. The idea behind all of these *nuanced signals* is that they are indirect.

DEFINITION

Nuanced signals are actions taken in unrelated circumstances that are nevertheless associated—or believed to be associated—with players of a certain type.

So far in this chapter I have described the following nuanced signals:

- Investing in expensive furnishings
- Increasing corporate dividends
- Participating in a house of worship
- Investing in one's new product launch
- Burning bridges
- Investing in new employees

All six strategies have two properties in common: all are costly signals, and all are actions that are only indirectly related to the "real" game being played. Other players in the game will infer something about the sender's type from these presumably associated behaviors.

Threat Strategies

Keep in mind that the whole point of studying signaling is to understand how one's actions communicate useful information to the other player. Holding a college degree is one kind of signal. Setting a price for an item to be sold is another. A simple and cheap way to send a signal is to just talk to the other player—but you already know that talk is cheap and is not likely to be reliable.

But what about sending a negative signal, say, a threat? In principle this doesn't sound any different, communication-wise. But what if the threat involved some sort of risk, something uncontrollable?

Brinkmanship

Brinkmanship relies on an equally nuanced signal as costly commitment goes, but using sticks instead of carrots. The idea is that one player is able to rig the game somehow so that a wrong move on the part of the opponent just might set something terrible and unstoppable in motion. In other words, making the wrong move just might push both of you over the brink, so to speak. While the concept has been around for ages, the first exposition and analysis of brinkmanship was put forth by Nobel Laureate Thomas Schelling in 1960.

DEFINITION

Brinkmanship is the art of coercing an opponent by linking the strategies of both players to a risky move of nature.

Game theorists Barry Nalebuff and Avinash Dixit provide a good example of this strategy from the Cuban missile crisis of October 1962. The Soviets had begun to prepare nuclear missile launching sites in Cuba, and the Americans detected this action. President Kennedy then announced a naval blockade of Cuba in order to prevent the missiles from arriving.

The crux of the issue was this: what would happen if the Soviets tried to challenge the U.S. Navy? Let's navigate through the logic here. Suppose Kennedy had issued the following blunt threat: "if you try to run our blockade we will nuke you." Such a threat would not be credible, in part because the outcome—mutually assured destruction—would be absurdly disproportionate as compared to the Soviet action of trying to get through a blockade. Besides, the United States would not even have been in any immediate danger. More likely, Kennedy would have played wait-and-see, which the Soviet leader, Khrushchev, would already have anticipated.

So how can Kennedy credibly create some grave risk for Khrushchev? He can't warn Khrushchev that he will roll dice and press the button if he gets doubles. Khrushchev will see through that, too. (Kennedy rolls double threes, but then realizes he can't annihilate the world just because he rolled doubles!)

What seems to have happened is that Kennedy and his advisors realized there was another player in the game, one they could not exactly control: the U.S. Navy. Navy officers had a detailed protocol on how to run a blockade. Kennedy realized that although he was the Commander-in-Chief, he was not the commander in the field or at sea. The navy regulations on how to run a blockade were much riskier than Kennedy would have liked—but that made things riskier for Khrushchev also.

You can almost imagine Kennedy communicating to Khrushchev, "I'd like to work with you, but I can't control those crazy navy hombres." As we know, Khrushchev (perhaps understanding the independence of his own navy) didn't like the risk himself, and decided to pull back from the brink by bringing the missiles home.

The key to brinkmanship is to ensure the threat is credible. In Chapter 4 you saw how a "weak" Goliath cannot scare off Upstart in the Market Entry game by threatening to fight Upstart in the marketplace. The reason is that Upstart knows that the weak Goliath will reap a higher return by accommodating rather than fighting.

Many people get carried away in their interpersonal dealings, especially in family situations, and make threats that cannot be acted upon. How many parents have threatened their children with dire punishment—"If you don't stop that, I'll never bring you to the playground again." Even small children know they can keep up their defiance in the face of a bogus threat.

On the other hand, what worked for President Kennedy is that his lack of control of the U.S. Navy was a believable, and worrisome, state of affairs. It's one thing to size up rational opponents who would not press the button, but it's another thing to invite randomness into the equation.

There's one more subtlety to the logic of brinkmanship. President Kennedy was truly unable to control the U.S. Navy because it was a separate entity. In fact, it is the very aspect of separability that makes such threats believable.

POISONED PAWN

You already learned that it's sometimes necessary to use mixed strategies in order to play your best. But do not confuse the carrying out of a properly computed mixed strategy in a well-defined game with threatening to carry out random play in other situations. Just because mixed strategies involve random play does not mean you can successfully threaten an opponent with arbitrary random actions. If the adversary understands that your random actions are not in your best interests, they will not be duly coerced. A random element alone does not make a threat credible.

A more modern day, and less frightening, illustration of this principle occurred in 2009 at the Copenhagen climate talks. The negotiators from the United States and China had reached an impasse. The Americans were pushing for outside verification of emissions reductions while the Chinese, although having agreed to make significant reductions, refused. President Obama pressed the case but was unable to make progress. As it turned out, yet again an independent entity was expected to make the difference: the U.S. Congress.

An energy bill in the House of Representatives had been passed with a provision to impose tariffs on imports from nations that did not lower emissions; and a number of senators threatened not to ratify any treaty that did not call for nations with competing goods to meet emissions targets. As it was reported in the media, China was more worried about the threat from these independent "loose cannons" than anything else. Once again, the right brinkmanship strategy involved an uncontrollable, somewhat random, element.

The Utility of Erratic Behavior

People like dependability. Many people, when faced with a choice to dine at either a chain restaurant (think McDonald's or T.G.I. Friday's) or an unknown local joint will opt for the chain restaurant. They know what they will be getting with the chain restaurant, whereas the local joint might be good, but it might be awful. Most restaurant chains understand this and work very hard to keep the food and atmosphere uniform among their different locations.

When people are dependable it has the same comforting effect on others. You know how they're going to behave. While it is clear that dependable behavior can enhance one's reputation for positive behavior, what can be said about erratic behavior?

Imagine being in a situation where you might be the target of an attack. Perhaps you are riding the subway or metro late at night and are afraid that the hooligans over there might assault you. What kind of signal can act as an effective deterrent?

Certainly, you cannot call out to your potential assailants, "Excuse me guys, please don't pick on me." One thing that might occur to you is to try to look tough. But how does the reader of a game theory book manage to look tough? One thing you can do is to put your hand in your pocket. This might successfully affect the attitude of someone readying themselves with a hidden weapon.

Another thing you can try is to act like someone capable of extreme behavior. If you sit there and mumble to yourself, or manage to put on a sort of crazed look, your potential assailants might seek a different target. Why bother with someone who might fight back, especially in some sort of unpredictable way? Both of these strategies—the hand in the pocket, and the crazy mumbling—depend on introducing a volatility and risk that might redirect your assailants to a softer target.

Veiled Strategies

Putting your hand in your pocket to feign a ready weapon might be an effective deterrent against criminals, but won't win awards for subtlety. But certain behaviors, it turns out, are so veiled that even the protagonists may not know what they are really signaling to the other party. These behaviors are not what they seem.

Bearing Outsized Gifts

At holiday time we often remind ourselves of the biblical verse, "it is better to give than to receive." What exactly are we signaling when we bestow gifts upon others?

> **NO BLUFFING**
>
> Recent research shows that it is indeed better to give than to receive. Scientific studies have shown that giving to charity increases our feelings of happiness, while spending money on ourselves does not.

A number of indigenous tribes in the Pacific Northwest still practice an age-old ceremony called potlatch, in which a feast is held for guests of a family or tribe. Among the many activities that might take place is the giving of gifts. While the purpose of the gifting is often for redistribution of wealth, there is an implicit understanding that hosting a potlatch, and especially the giving of lavish gifts, is a way of raising one's status.

As soon as we think of status, the idea of "keeping up with the Joneses" materializes, and it turns out that the potlatch ceremony can become a veritable status battleground. Stories abound of tribes engaging in back-and-forth gift-giving as a way of signaling superiority in no uncertain terms. One tribe would send boatloads of candle oil, copper sheets, and the like to another tribe, and the recipients, duly insulted, would send twice as much of their own stuff right back.

Some tribes burned up copious amounts of candle oil and other goodies at a potlatch, and such destructive behavior led to the entire practice being outlawed by both the Canadian and U.S. governments in the late nineteenth century. Apparently government officials wanted to put an end to wasteful and blatantly "uncivilized" behavior.

Now hold on there—generosity anywhere in the world can often have a subtext of status signaling. We (Westerners), for example, engage in a practice where our wealthiest citizens donate many millions of dollars to a university to be able to memorialize themselves while supporting an honorable activity like cancer research. It would be interesting to discover what proportion of such magnanimous contributions are given anonymously.

For many gifts, it may be very difficult indeed to tease out the truly good and generous intentions from the more manipulative ones. But when you consider the excesses of the potlatch ceremonies and the excesses of charitable contributions, once again the handicap principle comes to mind. Think again about extravagance: didn't you intend to show off, at least subconsciously, when you gave that big donation?

Losing Your Cool

Even the most even-keeled among us loses our temper occasionally, and often we feel intense regret immediately afterward. As consolation, we tell ourselves that no one's perfect and, depending on our angry deed, we might try to make amends and limit the damage.

But have you ever considered the possibility that your anger and fury might be the result of evolutionary hardwiring? I don't just mean that you inherited your hot-headedness from a certain side of the family. I mean that perhaps the tendency to occasionally lose it might have conferred some adaptive advantages to your ancestors.

Economist Robert H. Frank pioneered the discussion of how people's emotions might well play a strategic role in our interactions. Here's the idea. Consider an action, say one of revenge, that you might engage in. Although you might threaten to retaliate against your adversary, your rational side would calculate, in some sort of cost-benefit way, that following through with this act of vengeance is not worth it.

> **BET ON IT**
>
> Anger is the key emotion in our focus on whether someone will exact revenge. But other emotions also drive our behavior. Letting things like guilt, disgust, and love direct our actions can also be signals to our adversaries that we do not always behave according to what would rationally be in our best interests.

Similarly, the adversary is able to do the same calculation and would conclude that your threat of revenge is not credible. However—and this is the crux of the matter—these calculations are only valid for rational decision makers. As it turns out, suppose your thirst for revenge is uncontrollable and you will carry it out despite the fact that it is not in your best interest. Now here's the point: if the opponent knows that you are thus committed to acting on your emotions, they will avoid putting you in that position to begin with. Remember, when you carry out your retaliation, it has negative consequences for you, but it is harmful to them as well.

Let's retrace this argument in the following sequence:

1. You are predisposed toward emotional behavior.

2. An adversary is considering taking advantage of you.

3. If they take advantage of you, your rational response is not to retaliate because retaliating leaves you worse off than if you did not.

4. But your predisposition for emotional behavior means you are committed to acting rashly rather than rationally. Therefore you will likely fly off the handle and retaliate.

5. Your retaliatory act leaves you worse off but is bad for the adversary, too.

6. Your adversary decides it is too risky to cross you in the first place.

The upshot is that building a reputation for losing your cool can have long-term advantages. If you have a reputation for hotheaded retaliation, people will not take advantage of you in the first place. As Frank points out, the irony here is that an emotional bent, which makes someone unable to pursue their best interests, can in fact be advantageous after all.

The lesson, then, is that if a reputation for losing your cool precedes you, it can actually be a pretty effective bargaining chip.

The Least You Need to Know

- Nuanced signals are actions that are only indirectly associated with types and strategies in the actual game.
- Just as with more direct signals, nuanced signals still need to be costly to be effective.
- Tying a strategy to a random and risky action can be an effective way of generating a credible threat.
- The prospect of erratic behavior in an adversary can undermine analysis that was based on rationality.
- Demonstrating a history of emotional decision-making can serve to underpin a threat strategy.

Part 3
Getting Ahead by Working Together

Up to this point the focus has been on players as antagonists. Even in mixed-motive games like Stag Hunt, where there is an element of teamwork as well as competition, each player is basically working alone.

Now let's look into games of a more explicit cooperative nature. For example, when players team up in an organization, their synergy generates higher profits than the sum of what they can achieve separately. How should the players share this surplus profit, when each wants as much as they can get?

Other situations have the same cooperative, yet competitive, feature. When players bargain over an item's price, they each benefit but try not to give too much away. Same thing when an estate is divided up. Voting is another circumstance where individual's preferences may conflict as they work together toward selecting the best overall option.

The key to such games of cooperation is to find solutions that treat the players fairly. This is where game theory comes in. In this part I show you a variety of methods that deal with these problems extremely effectively.

An Overview of Cooperation

Chapter 9

In This Chapter

- How game theory treats bargaining
- Dividing up a set of items
- The synergy of cooperation
- Why voting is a game

Up to this point our study of game theory has focused on competition between two players. Reflecting on this, a couple of natural questions arise. First, is it possible to successfully study strategic behavior among many players? I mentioned earlier that if we consider more than two players, we'd have to worry about some of them ganging up on others. This leads to the second question: is some strategic behavior actually cooperative to some degree and, if so, how do we study it?

This chapter introduces the basic cooperative models in game theory. In these games the players still act in their best interests, but they benefit from cooperating with others. I will develop this theme by first looking at two player games with some limited cooperation, and then analyzing games with an increasing number of players and an expanding sphere of cooperation.

Bargaining with Two Players

Most of the game models I've described so far not only fixed the number of players at two, but treated just a small number of strategies for each player. I then introduced an element of communication between the players, called signaling, but reliable signals generally have to be costly. The other element of communication—direct talk—isn't very effective because talking about one's strategies in advance isn't likely to be convincing to the other player.

There is, however, an important realm of direct human interaction where two people are trying to enact a deal that will be good for both of them. But the way they interact is to discuss the outcomes, not the strategies, directly. When the outcomes are directly addressed in a strictly give-and-take way, we call it bargaining.

The natural way to set up a bargaining game is to use the extensive form. The merchant first sets a price; the customer comes back with a counteroffer; the merchant might lower the price ("for you"); the customer might then make a higher offer; and so on. Such a process is shown in the following figure.

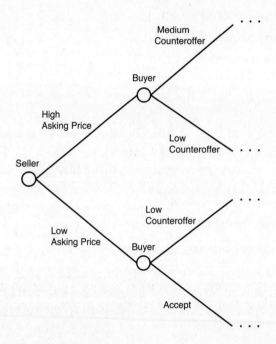

A bargaining game in extensive form.

While it is certainly possible to carry out some analysis using this model, the more popular and widely used approach is to conceptualize the two bargainers as players who are trying to narrow a monetary gap between them. If they are to close the gap at all—meaning they come to an agreement—there must have been some sort of surplus to share all along. Our approach is to find an acceptable way to share that surplus.

For example, suppose I have an object that I would be happy to sell for anything over $50, and suppose you would be willing to purchase this item for anything under $75. The surplus, or economic benefit here, is the difference between my lowest selling price ($50) and your highest buying price ($75). Looking at it this way, we have $25 between us that we will try to share. If we can't cooperate and reach an agreement, we each leave with nothing.

Looking at the topic of bargaining from this angle, the two players, rather than being entirely confrontational, are working together toward a goal. The way that John Nash modeled the problem was to consider the sum that they are working to share ($25 in our example), graph all of the possible outcomes, and come up with an acceptable formula to provide an allocation to the two players. The bargaining graph of feasible ways to share $25 is included here.

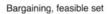
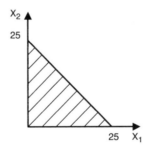

Bargaining graph of feasible outcomes.

Nash's approach, then, boils down to coming up with a fair way to divide the money. A "fair" division rule must satisfy a number of conditions intended to be reasonable to all. If the conditions are universally accepted, and a formula exists that satisfies them, then the two players would be compelled to accept the outcome as given by the formula.

Fair Division Problems

Nash's formula and other similar approaches work well when the situation involves just two people bargaining over money, which can be divided up in a lot of different ways. But what if the bargaining situation involves more than two people—for example, when an estate is divided up among several family members? And what if some of the items being fought over are objects that are not divisible? These aspects seem to require a broader theory.

What is usually termed fair division involves problems with two or more people, and often with a number of items, including indivisible ones, to be dished out. As with two-player bargaining, the challenge for game theorists is to come up with a fair allocation.

In fair division problems the players, as in the two-person bargaining scenario, still vie for shares of some predetermined goodies and are very much acting in their own self-interest. But fair division problems typically involve assorted items to be split up, and this means that different players often prefer different things.

Because of the varied preferences, fair division problems have less of a constant sum flavor than two-person bargaining does. So despite the fact that the players are still in conflict with one another, the context is more cooperative than when two people bargain over a fixed pie.

Let's get a taste of what these problems involve. Suppose a baby grand piano is part of an estate that is being divided among three siblings. The siblings might love one another dearly, but none of them wants to get the short straw.

Suppose that Susan really wants the piano, Marie wouldn't mind having it, and John has no use for it at all. We can imagine an arbitrator—or a trio of attorneys—who would interview the three siblings. Maybe one approach they could use is to convert indivisible items, like the baby grand, to monetary figures. The arbitrator might find out that Susan values the piano at $5,000, Marie at $1,250, and John at $0.

What we would have to do next is either award Susan the piano and arrange for her to compensate Marie and, possibly, John; or, if there are other items to be parceled out, instead of treating each one separately we might want to consider them all at once. There is another aspect to this process, though, that we shouldn't overlook.

Imagine that you are John. Why let Susan or Marie get a piano while you get practically nothing? You realize that if you got your hands on that piano, you could sell it on eBay for a lot of money. So John, while he harbors no ill will toward his sisters, might be inclined to treat his valuations strategically. Of course, his sisters might get the same idea, and such misrepresentation is something we will have to reckon with.

POISONED PAWN

Dealing with players misrepresenting their preferences in fair division problems is intriguing. While keeping players honest is an admirable objective, don't let it dominate your attempts to discover workable and useful apportionment techniques.

As with two-player bargaining, game theorists try to find solutions to fair division problems that satisfy certain reasonable conditions. Imagine that you and others go through a process where you split up a number of items—not just the baby grand, but some furniture, some electronic gadgets, and some miscellaneous stuff. When the process is complete you can't help but compare your allocation to the other ones. Do you prefer any of the other allocations to your own? If you don't envy the others' allocations, you have to be reasonably satisfied.

If there is a way to allocate the contested items so that every player always prefers their own share to all the other ones, the procedure is called "envy-free." While there are other desirable conditions we would want a division rule to satisfy, providing an envy-free solution might be the most important objective.

Work on fair division ranges from problems of cutting up cake to cutting up the Balkans, and the techniques that are out there range from absurdly simple to extremely convoluted. As with two-player bargaining, the best approach is not to let the players haggle in a free-for-all manner, but rather to elicit actions and evaluations from the players in a controlled way that results in allocations that are guaranteed to live up to certain criteria.

Cooperative Games

In Chapter 1 I asked you to consider a situation in which you and your three roommates sign up for a cable TV package that costs $86.17 a month. One easy way to divide up the charge is to split it up evenly among the four of you. But perhaps the bill would be much lower if a couple of you didn't insist on getting certain premium channels. The question is, what's a fair way to allocate the cost?

This problem of allocating the cable bill sounds similar to splitting up the bill at a restaurant. You might think it is a simple matter of finding out who ordered what dishes—or channels—and simply parceling out the relevant costs.

Let's back up a moment, though, and review exactly what is cooperative about such situations. Consider the restaurant bill. If two of you decide to share a dessert, you each pay half but you each only get half to eat. With the TV billing, once you have those premium channels, you can each consume as much TV as you want, at least within the 24 hours each day brings. It's not the same as consuming (half) the food at the restaurant!

To get to the heart of the issue, let's consider a more real-world problem. Suppose there are three departments in your company that are going to pay for the cost of super high-speed Internet access. A super high-speed network costs $1,000 a month; a fast network costs $700 a month; and a slower network (think DSL) costs $500 a month.

But suppose the different departments have different data requirements and therefore do not all need the same technology. Department A uses only relatively small files and needs just the basic network. Department B only needs the intermediate (fast) network. But Department C staff routinely download huge files and must have super high-speed access.

Because of Department C, your company is going to get the super high-speed network and cough up the $1,000 a month to pay for it. But given the different departmental requirements, what is a fair way to allocate the $1,000 per month cost?

Notice that since the three departments are united in one company, they need only one network instead of three. If the three departments separately arranged their own Internet access, they would pay a total of $1,000 + $700 + $500 = $2,200, but by virtue of their organizational union they only pay $1,000. This fact clearly shows the benefits of cooperating—in this case, by being billed together instead of separately.

But don't forget that there is still an element of competition here. The departments have separate, limited budgets, and no department wants to shell out more than it needs to for Internet access. While each department will reap the overall benefit of cooperating by sharing the single network, they are still fighting to pay as little as possible.

Let's invoke the concept of fairness when trying to find a solution to the game. We won't really have to consider the players as strategic agents, and in this setup we don't even have to consider them as bargaining with one another. We will, however, adopt an *axiomatic approach*, similar to the way Nash approached two-person bargaining, to reach a fair solution.

DEFINITION

An **axiomatic approach** involves setting down some principles that any fair (or "reasonable") solution ought to satisfy and then finding a solution that works.

Incidentally, it is in this type of cooperative model where ganging up on one another comes into play. Different subsets of players, called coalitions, are considered potential partners who "play" against the rest of the bunch. I put "play" in quotes because what we are actually doing is finding out how much profit or cost each such coalition would be able to generate, or would deserve to pay, if they operated without the other players in the original group.

Voting and Game Theory

That word *coalition* might elicit thoughts of voting. You may not have anticipated the topic of voting in a game theory book, but there are a number of fascinating aspects that certainly do belong here (and are explored more fully in Chapter 13).

First, why is voting a cooperative enterprise? The answer only requires us to think about what it means to make decisions in a democracy. It is through voting that the citizens come together and cooperate in the most elemental way: to jointly elect their representatives and leaders.

What about strategic behavior? We don't have to look too far to see how it can turn up. First, let's consider the two-party system that is, essentially, what exists in the United States. In most presidential elections there is a third party candidate (and others who are still more marginal). Sometimes this person has little effect on the race between the two main candidates, but sometimes the third party candidate makes all the difference in the world.

In 1992 Ross Perot played a major role in the presidential election. Perot garnered 19 percent of the popular vote, and we can certainly speculate that the Bill Clinton–George H. W. Bush race would have been pretty close without him. In 2000 Ralph Nader almost surely proved to be a "spoiler." Nader received less than 3 percent of the popular vote but, given the belief that most Nader supporters would have preferred Al Gore to George W. Bush, and given that Gore lost Florida—on which the entire election hinged—by only 537 votes, it seems safe to say that Nader's presence in the race made all the difference in the outcome.

The strategic element surrounding a third party candidate is that many people would not want to "waste" a vote on someone whom they thought had no chance to win. Indeed, it is possible that far more than 19 percent of the voters actually preferred Perot in 1992, or more than 3 percent preferred Nader in 2000, but such voters cast their ballots for someone else—presumably, their second-most-preferred candidates.

In light of this, you might suggest a different system altogether: with (say) three candidates, let each voter give their first choice three points, their second choice two points, and their third choice one point. Everyone votes, we add up all the points, and the winner is the candidate with the most points! Could such a system overcome the problem of voters not wasting their ballots?

There's one more cooperative aspect that I should touch upon here, and this is the issue of voters or their representatives getting together and cooperating in what are called blocs. Numerous countries have multi-party political systems very unlike the two-party system in the United States. Generally, with many political parties, no single party will gain a strict majority of the votes. Instead, to get a ruling majority, a number of parties must form coalitions. Similar behavior can take place among corporate shareholders, and very often the process of coalition-building creates some very strange bedfellows.

As you can see, the subject of voting is a rich one crying out for game-theoretic analysis. Let's summarize the main points thus far:

- Voting, although an individual act, is a cooperative exercise.
- Voting can involve strategic behavior.
- Some voting methods use a weighted point-counting system.
- Forming coalitions is often necessary to enact majorities in governments and corporations.

There are two other voting issues you need to consider. One is the innocent-sounding term, *proportional representation*. Every 10 years the United States (as well as other nations) conducts a census and, based on the census numbers, each state receives a number of electoral votes (which are used in the presidential election) that is proportional to that state's population share.

The electoral college allocates a fixed 538 electoral votes among the 50 states. Each state gets 2 votes for its 2 senators. Having accounted for the 100 Senate votes and the 3 for Washington, D.C., the remaining electoral votes apportionment should be the number of votes out of 435 that corresponds to each state's proportion of the U.S. population.

But what happens when a state like Pennsylvania, say, would be allocated 19.6 electoral votes based on its 4 percent share of the U.S. population? Since the electoral allocation must be a whole number, will Pennsylvania receive 19 or 20 votes? Virtually all states will have such fractional values, and it is not a trivial matter to decide how to round the fractions up or down and keep the total at 538.

Not all voting is confined to political races. Corporate politics often involve power struggles that accompany the voting process, and often the votes that political or corporate entities bring to the table do not carry the "power" that one would expect. Consider the following voting situation. A company has four shareholders:

- Shareholder 1 owns 100 shares
- Shareholder 2 owns 100 shares
- Shareholder 3 owns 80 shares
- Shareholder 4 owns 50 shares

First notice that the total number of shares is 330. Suppose that the shareholders only enact policy changes when a majority of 170 shares is reached.

The question for now is, does Shareholder 4 have any real say in corporate governance? The way to determine this is to find out whether Shareholder 4 is a necessary member of any winning coalition of shareholders. In other words, do any other combinations of shareholders need Shareholder 4 to join them in order to enact policy changes? It turns out that the answer is no.

While you might shrug your shoulders at this quirky turn of events, notice that Shareholder 4 owns 50 out of 330 shares, which is over 15 percent of the total. It doesn't seem fair that someone who owns 15 percent of the shares has no say at all in a system that otherwise appears to be democratic. Such paradoxes play a role in determining, for example, the "power" that various states wield in the U.S. electoral college. In Chapter 13 you will learn more about strategic issues in voting, such as how to measure power among different constituents or blocs.

The Least You Need to Know

- Two-person bargaining is about sharing a surplus.
- Fair division problems have multiple items that players value differently.
- Cooperative games involve players who join in productive coalitions.
- Voting involves both cooperative and strategic behavior that can be investigated using game theory.
- All of the cooperative models revolve around conditions that capture fairness and solutions designed to meet the conditions.

Bargaining Games

Chapter 10

In This Chapter

- Two-player bargaining
- The axiomatic approach
- Nash's solution
- The Kalai-Smorodinsky alternative

This chapter considers the problem of two players bargaining over how to divide a sum of money, given that they may have certain legitimate claims. Rather than employ a back-and-forth approach to bargaining—with dramatic offers and counteroffers, the following procedure is like calling in an impartial arbitrator to determine a fair and final allocation.

Nash's Standard Model

The way John Nash considered the problem of two-player bargaining was to suppose there is a set of different outcomes represented by coordinate points (u_1, u_2) that are available to the players. Even though the players are bargaining over money, the outcomes can be measured more generally as utilities (hence the u notation). The players need to agree on one of the joint outcomes or else they both end up at an inferior *disagreement point*.

Instead of modeling a back-and-forth "haggling" process where two players are trying to narrow a monetary gap between them, let's represent the bargaining problem as one of sharing a sum M of money. This sum could be the amount of money "on the table" in a negotiation, the amount left over in an estate, or the total assets of a bankrupt firm. One source of friction is that the players might have legitimate claims that can't both be satisfied. Another source of concern is that the players might value monetary payments very differently; this can happen, for example, if one player is rich and the other is poor.

> **DEFINITION**
>
> The **disagreement point** is an inferior outcome which results if the players can't agree on the final outcome or disbursement. In our treatment, the disagreement point is (0, 0). This means that neither player gets anything out of the bargain. Using (0, 0) is a slightly less general approach than what Nash proposed, but it allows us to avoid some technicalities.

Even though the two players have diametrically opposed interests, the game is cooperative in two very important ways. For one thing, the threat of being stuck with the disagreement point is an incentive for the players to work together.

> **BET ON IT**
>
> The disagreement point (0, 0) is the maximin point for each player. In other words, by using (0, 0) as the disagreement point, either player can walk away from the negotiation with a guarantee that by exiting they neither gain nor lose anything.

But the real driver behind Nash's approach is to get the two players to agree on a set of rules that an allocation procedure should satisfy. If rules can be devised that are universally acceptable, and if there is a mathematical way to obtain an allocation that satisfies them, then we would have a ready-made allocation scheme.

Nash was not only able to provide a set of conditions that most people find to be reasonable, but he also found the only mathematical solution that satisfies them. So if any two players agreed to the proposed rules, they would logically have to accept the corresponding solution—which explains why this approach is like using an impartial arbitrator.

I will fill you in on Nash's reasonable conditions in a moment, but first I need to complete the description of the bargaining game.

The bargaining game is set up as follows:

- There are two players.
- Set S is the set of feasible outcomes.

- Outcomes for the two players are denoted by (u_1, u_2), where Player 1 obtains utility u_1 and Player 2 obtains utility u_2.
- If players can't agree on an outcome, they end up at $(0, 0)$.
- The bargaining solution will be given by the single point (u_1^*, u_2^*) that satisfies the conditions N1–N6 (explained in the next section).

To illustrate this game, suppose that two players who value money equally are to bargain over $10,000. The set of points S they are bargaining over is given by the triangular region in the following diagram. These are the points (u_1, u_2) such that $u_1 + u_2$ is no more than 10,000. Points on the diagonal line represent payoffs (u_1, u_2) that add to exactly $10,000.

It is sensible to ask under what circumstances the two players would walk away with a total of less than $10,000. One answer is that they are "inefficient" bargainers. Another answer is that perhaps they have had some of the original sum put aside, whether for charity or for tax purposes, and so on. For such reasons, the feasible set S must include the points below the diagonal boundary line, too.

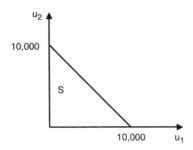

Bargaining equally.

Now suppose that Player 1 can produce a legitimate document that shows he is owed $6,000. One way to interpret such a claim is to say that this player is not owed anything more than $6,000. Observe how, in the following graph, the feasible region has all coordinate points with u_1 greater than $6,000 (the shaded area) eliminated from consideration.

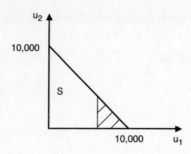

Bargaining with one claim.

Some Reasonable Assumptions

Nash's approach to the bargaining problem is axiomatic. In other words, he set down incontrovertible principles that a bargaining solution must satisfy, and then found the solution. Nash's axioms are labeled N1 through N6:

N1. **Individual rationality:** $(u_1^*, u_2^*) \geq (0,0)$.

> Condition N1 simply means that the bargaining solution must be at least as good for each player as what they would get from walking away.

N2. **Feasibility:** (u_1^*, u_2^*) must be in the set S.

> Condition N2 states that we can't allocate something that's not there to dish out.

N3. **Pareto optimality:** This means that there can't exist any other proposed bargaining solution that dominates (u_1^*, u_2^*). In other words, once we find the Nash solution, we can rest assured there is no other solution out there that is at least as good for both players and strictly better for one of them.

> **BET ON IT**
>
> Vilfredo Pareto, a turn-of-the-last-century economist, wanted to establish a social welfare criterion whereby if a change in wealth is made in a society, no one gains at someone else's expense. A Pareto improvement in wealth is an outcome in which no one is worse off than before. In the bargaining game, the Pareto optimal outcomes are the ones where Pareto improvements are no longer possible. These points lie exclusively along the outer boundary of S, also called the "efficient frontier" by economists.

N4. **Independence of irrelevant alternatives:** Let R be a subset of S (that is, R lies completely inside of S). If the solution (u_1^*, u_2^*) for S is a point in R, then it must be the solution to the bargaining game defined for the smaller set R.

Another way to view this property is that if "irrelevant alternatives" are deleted from the initial set of possibilities, the solution will not change. And yet a further way to interpret this property is to say that if the feasible set is enlarged, the solution to the larger game is either the same point or one in the added set, not a different point in the original set.

It is helpful to see this visually. In the following graph, the shaded region is "irrelevant" since this region can be deleted without changing the indicated solution point.

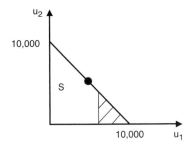

Irrelevant alternatives.

N5. **Independence of linear transformations:** For our purposes this principle means that if you converted your currency from dollars to euros, for example, then the relative allocations will not change.

N6. **Symmetry:** The symmetry principle says that if S is a symmetric set—meaning that the "mirror image" of every point (u_1, u_2) in S is also in S—then u_1^* must equal u_2^*. Put differently, if there are two players bargaining over a sum of money such that they make identical demands and have exactly the same utility for money, then they must be treated equally, meaning that they must be awarded the same payoffs.

>
> **POISONED PAWN**
>
> Although reasonable at first sight, the principle of the independence of irrelevant alternatives is the most controversial of Nash's axioms. With this principle, Nash tells us that because the points in the irrelevant subset are less preferred than the solution point, then discarding them should make no difference. (Discarding them reduces the original set S down to the subset R.) In a number of contexts outside of bargaining, however, this principle does not seem to apply.
>
> For example, think of a three-way political race. Suppose Candidate A is ahead of B in the polls, and C trails behind both. If C drops out of the race, Candidate B might pull ahead of A, because it is possible that C's supporters now back B. In this case C is certainly not an irrelevant alternative.
>
> Despite this criticism (which after all is in a voting, not bargaining, context), if we stick to the given bargaining situation the independence of irrelevant alternatives principle is defensible.

Nash proved that there exists exactly one solution that satisfies conditions N1 through N6. Let's look at the solution and examine its behavior.

The Nash Solution

The Nash solution is simple: the sought-after solution (u_1^*, u_2^*) is the point in S that maximizes the product u_1 times u_2. This point lies on the boundary of S, and it has a nice geometrical interpretation: it is the point that maximizes the area of an inscribed rectangle.

Application to Bankruptcy Problems

Let's see how the Nash solution works in the bargaining context. Consider a situation where a firm goes bankrupt and it owes two creditors money.

Bankrupt firm A owes Player 1 $50,000 and Player 2 $80,000. However, firm A only has $60,000 left. What is the Nash solution to this bargaining game?

First, the bargaining set is illustrated in the following graph:

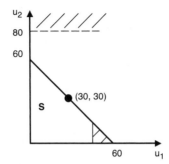

Payoffs are in thousands of dollars.

The bargaining set in the Bankrupty game.

The fact that the firm owes Player 2 $80,000 is displayed by the dashed line in the graph. Since there is only a total of $60,000 available, the claim of $80,000 does not affect the feasible region, since the only points that are discarded already lie outside S. However, since the firm owes Player 1 $50,000, we eliminate all points with u_1 coordinates above $50,000 (these are in the shaded area). Simply looking at the graph, we can see that the Nash solution—finding the boundary point yielding the largest rectangle in area—is to give the players $30,000 each. Notice how the set of points with u_1 coordinates above $50,000 is a set of irrelevant alternatives.

NO BLUFFING

Similar problems involving the carving up of an estate, along with some provocative solutions, were described in the Talmud, the ancient book of Jewish law, many centuries ago. These problems were rediscovered in the 1980s, prompting a lively debate among game theorists. It turns out that some of the mysterious Talmudic prescriptions actually coincided with certain game-theoretic constructs!

It might strike you as odd that each player receives the same sum despite their very different claims. On the other hand, if you saw no reason to disagree with the six Nash conditions, you would have to accept the allocation.

Now suppose a new fact is discovered: a mistake has been made and Player 1 in fact is owed only $20,000, not $50,000. In this case the Nash bargaining solution will give Player 1 all $20,000 she is owed. It is a bit troubling that one player now receives her full claim, while the other player only gets half of his.

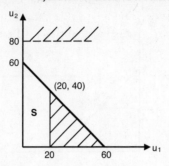

Modified Bankruptcy game.

Let's now think about how the players view the money, and not focus solely on the dollar amounts. Suppose that Player 1 has an insatiable appetite for money. No matter how much Player 1 has in the bank, every additional dollar is just as desirable. On the other hand, let's suppose that Player 2 is more easily satisfied with smaller sums. In other words, Player 1 is greedy and Player 2 is poor.

In this scenario, Player 1's appetite for money causes him to bargain more aggressively. But Player 2, on the other hand, has more modest desires and doesn't drive as hard a bargain as Player 1. Let's look at an example to see how this works. The amount that the two players are bargaining over is $1,000.

The best outcome for either player is to receive all $1,000. Now set up a scale measuring the utility for each player. Suppose the maximum utility, which comes from a payoff of $1,000, is measured as 1.0 for both players on your scale.

Player 1's appetite for money never changes; in other words, it increases at a constant rate. So for Player 1, $500 is worth five times as much as $100, and $1,000 is worth ten times as much. So a payoff of $100 has a corresponding utility of 0.1 on the scale; a payoff of $500 has a utility of 0.5, and so on, up to 1.0.

But Player 2, being more easily satisfied, gets a bigger kick out of smaller payoffs than Player 1 does. Player 2's utility from a payoff of $100 is 0.4 (much higher than the 0.1 for Player 1), and her utility for a payoff of $500 is 0.75.

The following table fills in more of these utilities for both players.

Payoff ($)	100	300	500	700	900	1000
Player 1	0.1	0.3	0.5	0.7	0.9	1.0
Player 2	0.4	0.6	0.75	0.86	0.95	1.0

Remember that the Nash bargaining solution finds the point that maximizes the product of the players' utilities. The next table shows five different ways to split up the $1,000, as well as the corresponding utilities for both players and the product of the utilities.

Solution point	Player 1 utility	Player 2 utility	Product
(100, 900)	0.1	0.95	0.095
(300, 700)	0.3	0.86	0.258
(500, 500)	0.5	0.75	0.375
(700, 300)	0.7	0.60	0.420
(900, 100)	0.9	0.40	0.360

Looking at the last column, you see that the highest product of the two players' utilities occurs at (700, 300). (Of course, the table does not show all of the many other possible ways to divide the $1,000, and therefore it is possible that the exact solution differs somewhat from (700, 300).)

This reflects the ability of the richer player to "bargain down" the poorer player (although as we know, no actual bargaining interaction is taking place).

When players 1 and 2 reveal their types, as it were, the corresponding utility measurements determine the lopsided outcome. Paradoxically, the poor individual gets a smaller share, and for many of us this might be a troubling feature of this particular solution. In Chapter 18 you will learn much more about assigning utilities to outcomes in games.

The Kalai-Smorodinsky Solution

Seeing as the Nash solution to the Bankruptcy game examples does not appear to take some important information into account, game theorists have tried to develop other solutions that are more appealing.

The main focus of the newer approaches has been to single out condition N4— independence of irrelevant alternatives—as the troublemaker, and replace it with a different, less contentious, axiom. The best-known of the alternative solutions was published by Ehud Kalai and Meir Smorodinsky in 1975. They noticed that it was possible in some bargaining games for the feasible set S to expand and yet for the Nash allocation to one of the players to decrease.

Kalai and Smorodinsky sought to remove the irrelevant alternatives axiom and instead use a condition that would avoid N4's negative ramifications. The condition they used directly addresses the problem of how to make sure solution outcomes do not plummet when the feasible set grows. Their replacement axiom is called *monotonicity*.

> **DEFINITION**
>
> **Monotonicity** states that if (u_1^*, u_2^*) is the solution for set S, and if S is a subset of a larger set W, then the solution outcome in W will dominate (u_1^*, u_2^*). In other words, if the players are bargaining over a larger pie, monotonicity ensures that neither one will do worse.

Kalai and Smorodinsky showed how to obtain an alternative bargaining solution when the monotonicity rule replaces independence of irrelevant alternatives. Let's call their solution K-S. Let's look at the K-S solution and contrast it to Nash's solution.

The K-S solution is found with the following easy steps.

1. Establish the feasible set S.
2. Find the maximum utility that Player 1 can feasibly attain (without giving Player 1 more than is owed). Call this U_1-max.
3. Find the maximum utility that Player 2 can feasibly attain (again, without allocating more than is owed). Call this U_2-max.
4. In general, the point (U_1-max, U_2-max) will lie beyond S. Draw a line from (0, 0) to (U_1-max, U_2-max). The K-S bargaining solution is the point where this line crosses the boundary of S.

Let's revisit the first Bankruptcy game from before. The players are bargaining over $60,000. The firm owes Player 1 $80,000 and Player 2 $50,000. See the following figure for the K-S solution to this game.

The point (50, 60) in the diagram is the pair of maximally attainable payoffs for the two players, respectively. The firm owes Player $50,000. Since 60 is available, a payment of 50 is possible but nothing beyond that is owed. The firm owes Player 2 $80,000 but her maximum payoff is 60, since only 60 is available.

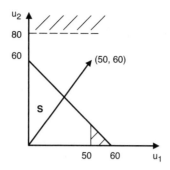

Bankruptcy game with the Kalai-Smorodinsky solution.

The intersection of the two lines occurs at the point (27.273, 32.727). So the Kalai-Smorodinsky solution will pay Player 1 $27,273 and Player 2 $32,727. Remember, the Nash solution was $30,000 each. K-S allocated more to the player who was owed more.

Now, what if the bankrupt firm A actually had $100,000 in assets instead of $60,000? The Nash solution gives $50,000 to each player. Notice how this pays off Player 1 completely, but not Player 2. The K-S solution pays $38,462 to Player 1 and $61,538 to Player 2. Neither player is paid off in full under K-S, but once again, the player who is owed more is paid more.

What about when the players have different utility for money? We saw that the Nash solution awarded more to the wealthier player. Unfortunately, the same is true of the K-S solution.

> **BET ON IT**
>
> A number of game theorists have developed bargaining games in extensive form where the sequential moves model offers and counteroffers. Although these models yield some nice results, they involve more difficult mathematics and do not have the same widespread applicability and appeal as the Nash model.

To summarize what you've seen, there are two well-established axiomatic approaches to the problem of bargaining over a monetary sum. If you believe irrelevant alternatives are truly irrelevant, you might prefer the Nash solution. If instead the monotonicity principle appeals to you, then the Kalai-Smorodinsky solution is the ticket.

There is a curious aspect to both solutions, however, which is that players who derive more utility from money—greedier players, if you like—will get a bigger award. While a sympathetic reader might find this characteristic to be unfortunate, it is simply a mathematical consequence of the information that the players provide. We should note, though, that in "real life" those who bargain more aggressively do generally come away with more. If it is true that the meek shall not inherit this particular world, game theory is simply the messenger.

The Least You Need to Know

- The game theory approach to bargaining is axiomatic: set down reasonable conditions and find a solution that obeys them.
- The Nash and Kalai-Smorodinsky solutions are tailored to slightly different sets of underlying principles.
- Unlike the Nash solution, the K-S solution never decreases a player's payoff when the pot is increased.
- Players who are risk averse—more cautious or timid—are awarded less than players that are aggressive or greedy.

Fair Division Games

Chapter 11

In This Chapter

- Cutting things evenly
- Dividing up many items
- Preventing envy
- Keeping portions equitable

Like bargaining problems, fair division problems typically involve giving away items that come out of nowhere: a parcel of land, for example, or valuables bequeathed from an estate.

However, in fair division problems, there are often more than two players. Some fair division problems involve stuff that is feasible to cut up into pieces or divide, such as money, while other problems involve things that can't reasonably be split at all, such as a piece of jewelry or a painting. When items are indivisible, giving them away is an all-or-nothing affair, and in these cases the players that come away empty-handed are often compensated with a monetary payment.

I Cut, You Choose, and Related Methods

The simplest example of a fair division problem usually involves two children sharing a piece of cake or some other goody that can be cut up into pieces. Bear in mind that the cake might not be uniform; in other words, one part might have more icing while another part might have more crust.

An ancient, King Solomon–like solution to this problem is known to many wise parents: let one child cut the cake into two pieces that she deems to be equally attractive, and let the other child choose first. The kid who does the cutting will try to make sure that either piece would be satisfactory; more precisely, to her eye both pieces would be as closely matched as possible. Even little kids understand that if the pieces they cut are patently uneven, the second child will get the better one.

One thing to keep in mind is that in fair division problems, just as in almost all of game theory, each player only cares about his or her own outcomes. So with two children dividing the cake, we must assume that neither kid cares what the other one ends up with. As we know, some kids (and grown-ups, too) indeed factor others' pleasure into their own utilities, but we will not consider such empathetic behavior just now.

BET ON IT

A lot of fair division literature examines cakes to be sliced up, and the cake-cutting is generally a metaphor for something important: carving up Berlin or the Balkans after World War II, or the ongoing problem of dividing up land in the Middle East.

Now that you are warmed up on fair division with two players, it's time to try a cake-cutting solution with three people. The so-called Steinhaus-Kuhn Lone Divider Procedure is one way to extend the "I cut, you choose" cake-cutting method to three players.

The Steinhaus-Kuhn Lone Divider Method

Suppose we have a cake to be divided up among three players. Further, let's assume that the three players follow the rules scrupulously. The steps of the procedure are as follows. A player thinks a certain piece is acceptable if it gives them at least a *proportional* share.

DEFINITION

A fair division method for *n* players is **proportional** if it guarantees that each player believes that she receives at least $1/n$ of the total prize. So if there are 3 players, the division is proportional if each player believes that she receives at least $1/3$ of the total.

The initial steps of the procedure are as follows:

1. Player 1 cuts the cake into three pieces he believes are equal in value.
2. Player 2 indicates which of the three pieces she believes are acceptable and which are not, provided that she must find at least one of them to be acceptable.
3. Player 3 also points out which pieces are acceptable or not, subject to the same proviso as Player 2.

Now things get a little complicated. Suppose that Player 2 or Player 3 (or both) found at least two of the three pieces to be acceptable. Let's say it is Player 2. We then let Player 3 pick first. (We know that Player 3 finds at least one piece to be acceptable, and we'll assume he picks one that he thus likes.) After that, we let Player 2 pick. (We know that Player 2 thought there were at least two acceptable pieces.) Finally, we let Player 1 pick. Player 1 did the cutting in the first place, and the assumption is that he cut them equally, at least to his eyes.

If Player 3 liked two of the pieces, and Player 2 only liked one piece, we let Player 2 pick first—now she's happy—and then Player 3 picks one of his acceptable options, and finally Player 1 is left with a piece he will think is decent.

Now suppose that Player 2 and Player 3 each liked just one of the pieces. If they prefer different pieces, that's great: we give them each the ones they liked and then Player 1 gets the remaining piece, which, once again, he won't mind.

But what if players 2 and 3 each prefer the same piece of cake? Here's what we do: randomly give Player 1 one of the two pieces that players 2 and 3 rejected. Since players 2 and 3 rejected the piece now going to Player 1, they believe they end up with a total of more than $2/3$ between them. Now we recombine the two remaining pieces (which includes the one they both favored) and players 2 and 3 play divide-and-choose between them. Since they are dividing more than $2/3$ of the cake (as they see it) they will be fine.

POISONED PAWN

The steps for dividing the piece of cake among three players are not quite airtight, but going through the final details of the argument is beyond the scope of this book and won't alter the procedure.

As convoluted as these steps may seem, it is actually possible to extend the logic of the procedure to cover instances of four players and even five. I will not give the details here, but it can be done.

As I pointed out in Chapter 9, one important measure of the success of a fair division method is whether the players end up wishing they had someone else's allocation. If all players like their own portions at least as much as any others, then we say the allocation is envy-free. Let's see if this condition applies to the Steinhaus-Kuhn Lone Divider Procedure.

One of the possible scenarios had Player 2 finding two pieces acceptable, in which case Player 3 would pick first. While Player 2 still ends up with an acceptable piece, it is possible that she would have preferred the piece that Player 3 made off with. So, in general, this procedure, while it guarantees allocations that are acceptable to all, does not guarantee an envy-free solution.

Another difficulty is that the procedure can be manipulated by the players. We just saw how Player 2 can end up envying Player 3 in the case where Player 2 identified two acceptable pieces. To pre-empt this outcome, Player 2 can instead identify just one acceptable piece—the one she really wants—and at least some of the time (when Player 3 picks a different one) she will come out better.

There are other ways the players can get around the rules. Given that the cake is not homogeneous, we can imagine that if Player 1 knows the preferences of the other players, he might be able to cut up three pieces so that one of them is perfect for him, but is likely to be rejected by the other two. And finally, who decides which players go first, second, or third? There seems to be some advantage in going third, and it would appear that making the ordering fair would depend on assigning a random sequence to the players.

Other Divisible Procedures

Theorists have devised other procedures to treat fair division problems with a single divisible item. One of them involves one player moving a knife over the cake and the next player calling "stop" and getting the piece that would be lopped off at that moment. Some procedures have two knives moving along in parallel. One notable method, called the Last Diminisher Procedure, involves the players taking turns cutting off larger slices and then trimming these pieces to get them down to the right size.

To summarize the main theme of these cut-and-choose procedures, the players themselves—not an outside arbitrator—carry out the dividing, which is done with the goal that all receive a piece that in their eyes is at least proportional. Envy-freeness, if it happens, is icing on the cake.

Proportionality and Envy-Free Allocation

Now let's look at situations in which a set of items is to be divided among players, but in these applications the items themselves must be kept whole. Instead of cake, or plots of land, we are now dealing with individual objects, such as family heirlooms that might be part of an estate that is being contested.

The Knaster-Steinhaus Procedure

In 1948 the Polish mathematician Hugo Steinhaus published the details of a procedure to fairly assign a number of contested items. He attributed the work to his colleague, Bronislaw Knaster. We will assume that each player can monetarily evaluate each of the items in the set to be divided up.

NO BLUFFING

The subject of fair division was initiated during World War II by mathematician Hugo Steinhaus and his colleagues Bronislaw Knaster and Stefan Banach, and published by Steinhaus in 1948. It was not initially a branch of game theory, but over time it has been game theorists, in particular Steven Brams and Alan Taylor, who have contributed much of the theory.

Let's describe the procedure. We have a set of players, each of whom looks over the multiple items to be divided up and assigns a monetary value to each one. Each valuation is independent of the other valuations; in other words, the fact that a player might be granted a certain item has no effect on the valuation of any other item. Based on the valuations, we first calculate a "fair share" to dish out to each player. Then we create an allocation by giving each item to the player who values it the most.

Doing this creates some winners and losers in the following sense: some players receive items with a much greater total value than other players. The players whose awards thus generate a surplus of value—value beyond their fair share—will make monetary payments to compensate those who end up with a deficit (value less than their fair share).

To see how the method works, let's carry out an example thoroughly. Here are the steps of the Knaster-Steinhaus Procedure:

1. Have each player write down how much each individual item is worth to them.
2. Add up the valuations and obtain a total valuation for each player.
3. Calculate an initial fair share for each player. Each player's initial fair share is just their total valuation divided by the number of players.
4. Assign each item to the player who valued it most. (Let's not worry about ties right now.)
5. Total up the valuation of the allocated items and compare each player's allocation to their initial fair share.
6. Record the difference between the allocated value and the initial fair share as either a surplus (when allocated value exceeds fair share) or a deficit (when fair share exceeds allocated value).
7. Calculate the total surplus by adding the terms in step 6 above. Then divide by the number of players to get the average surplus.
8. Find an adjusted fair share for each player by adding the average surplus to their initial fair share.
9. Determine a final settlement using the following conditions: (a.) each player whose allocated value exceeds their adjusted fair share will pay the difference; and (b.) each player whose allocated value is less than their adjusted fair share will receive the difference.

Let's walk through these steps with an example in which there are three players—Susan, Marie, and John—and four items—A, B, C, and D—to be allocated. Each item must remain whole. We create the following table and list the three players' valuations of the four items. Next, we compute the total valuations.

What drives the method is that the players, in general, will assign different valuations to the different items. Susan, for example, values item A at $12,000, but it is only worth $6,000 to Marie. The total valuations may widely differ, reflecting the players' very different preferences. Still, keep in mind that giving an item to the player who values it most seems utterly reasonable, and when some players thus lose out in assigned value, it seems only fair to compensate them by sharing the surplus wealth in a proportional manner.

Valuation	Susan	Marie	John
Item A	$12,000	$6,000	$8,000
Item B	5,000	1,250	1,000
Item C	1,000	500	2,500
Item D	750	3,000	500
Total Valuation	$18,750	$10,750	$12,000
Initial Fair Share	6,250	3,583	4,000
Items Received	A,B	D	C
Difference (= surplus/deficit)	17,000-6,250 = $10,750	3,000-3,583 = $-583	2,500-4,000 = $-1,500
Share of Surplus	= 2,889	= 2,889	= 2,889
Adjusted Fair Share	6,250 + 2,889 = $9,139	3,583 + 2,889 = $6,472	4,000 + 2,889 = $6,889
Final Settlement	A,B - $7,861	D + $3,472	C + $4,389

Knaster-Steinhaus example of dividing property among players.

To check the calculations, observe that the total valuation for each player is just the sum of the valuations in their column. For example, Marie's valuations add up to $10,750. Next, we find each player's initial fair share. Marie's is $10,750 divided by 3 (the number of players), which equals $3,583. The justification for this is that if Marie values the entire set of items at $10,750, then in her eyes each of the three players deserves an equal (third) share of that sum.

Now we assign each item to the person who values it the most. Marie valued item D more than Susan or John did, so we award D to Marie. However, since either Susan or John valued the other items more, item D is Marie's only prize. Since Marie valued D at $3,000 but her fair share was $3,583, Marie has a deficit of $583.

Susan racked up a huge surplus when she received items A and B. When Susan's surplus is combined with Marie's and John's deficits, the sum works out to $10,750 – $583 – $1,500 = $8,667. When this surplus is shared equally—i.e., proportionally—it works out to $2,889 each.

BET ON IT

It is important to realize that much of this surplus arises from assigning item A to Susan (she valued it at $12,000). More generally, this positive surplus occurs precisely because the players differ in their valuations. The sharing of the surplus is done simply by sharing the player's total perceived wealth equally.

To finish determining Marie's final settlement, we add the surplus of $2,889 to Marie's initial fair share amount of $3,583 and we obtain a value of $6,472. This is Marie's adjusted fair share: the original amount she thought was fair plus an equal share of the surplus wealth. Since Marie obtained D, which is worth $3,000 to her, she is still owed $3,472, which will be paid in cash. Susan and John's shares and settlements are computed similarly.

As I noted previously, the primary goal is to find a proportional procedure, and if we obtain an envy-free division, that's a bonus. In our particular example, the allocation is envy-free. Just to get a hint of what you need to check: suppose you are John. John obtains C (worth $2,500 to him) and $4,389 in cash for a total of $6,889.

When John looks at Marie's award, he sees that she got item D (only worth $500 to him) plus $3,472 in cash for a total of $3,972. So John certainly does not envy Marie. It is easy enough to check all of the other combinations to verify that the final settlement is envy-free for the other players as well.

Unfortunately, the Knaster Steinhaus method is not envy-free in general, although it is when there are only two players. Another drawback of the procedure is that to implement its results, the players each have to have a bankroll to enable them to make cash payments, if necessary. Finally, as with the lone divider method discussed in the previous section, the valuations in Knaster-Steinhaus can be manipulated by the players.

> **POISONED PAWN**
>
> The Knaster Steinhaus method is subject to players' strategically misrepresenting their valuations. For example, if Marie claimed that item A was worth far more than $6,000 (but less than $12,000), then her initial fair share would be much higher and her final settlement would provide her with a bigger cash payment than if she honestly reported her valuations.

Despite the criticisms, the Knaster-Steinhaus Procedure is lucid and provides a proportional fair division rule for indivisible items. On top of that, fair division experts Steven Brams and Alan Taylor point out that the rule is efficient, meaning in this context that the total valuation is maximized. This in turn means that we can't improve someone's outcome without making a different player worse off.

The Adjusted Winner Procedure

In the next two procedures we assume that the items in the estate can be shared, if not physically chopped up. For example, suppose one of the items is a vacation home that is to be shared 60-40. Instead of imagining that the home can be cut into two parts, we will treat the partial allocations as is done with time-share properties: one player uses the home 60 percent of each year, and the other player uses it 40 percent of the year. In this way we can get around the winner-take-all constraint that existed in the Knaster-Steinhaus Procedure.

Let's work through the Adjusted Winner Procedure, developed by Brams and Taylor. There is a set of m items to be divided between two players. Each player looks over the m items and allocates a total of 100 points among them. The points represent the relative preferences of the players for the various items. (More points means a greater preference.) For the moment, let's assume that the players' evaluations are done honestly.

Suppose Player 1 is Susan and Player 2 is Marie. Here are the steps we need to follow in the Adjusted Winner Procedure:

1. Find all the items that Susan valued more than Marie. Now add up all of Susan's points for those items.

2. Similarly find all items that Marie valued more than Susan, and add up all of Marie's points for those items.

   Let's suppose, without loss of generality, that Susan totaled up more points in step 1 than Marie did in step 2.

3. Create an initial assignment by giving Susan the items for which she "outpointed" Marie, and giving Marie the items for which she outpointed Susan. If there are any ties, start out by assigning those items to Susan.

4. Order the goods assigned to Susan as follows:

   (a.) Create ratios S_i/M_i, for each item i, of Susan's valuation to Marie's valuation. These ratios are always greater than or equal to 1 (since there was either a tie or Susan outpointed Marie on those items).

   (b.) List the item with the smallest ratio first, the item with the second smallest ratio second, and so on.

   Note that the item with the smallest ratio has the closest relative comparison in value, while the item with the largest ratio has the biggest relative discrepancy in value.

5. Since Susan has more points than Marie, to make the assignment more *equitable*, transfer items from Susan to Marie, starting with the smallest ratio items.

6. To achieve an equitable final allocation, we may need to transfer a fraction of an item. (This is where the assumption about fractional sharing comes into play.) The appropriate fraction is the one that brings both players' total number of points to the same level.

DEFINITION

An assignment of items to players is **equitable** if each player thinks that what they get is worth the same to them as what the other player gets as valued by the other player.

Let's work out the example that appears in Brams and Taylor's book. Susan and Marie are to divide three items called A, B, and C. They each allocate a total of 100 points over the three items.

	Susan's Valuations	Marie's Valuations
A	6	5
B	67	34
C	27	61

Adjusted Winner Procedure for allocating property.

First check that each player has allocated exactly 100 points among the items. Since Susan values A and B higher than Marie did, and Marie valued C higher than Susan, we initially assign A and B to Susan and C to Marie.

Among Susan's items:

- Item A has a ratio of 6/5 = 1.2
- Item B has a ratio of 67/34 = 1.97

Susan has valued items A and B (initially assigned to her) at 6 + 67 = 73 points, while Marie values item C at 61 points. Since Susan has a higher valuation than Marie, the initial assignment is not equitable.

The item with the smallest ratio is A, so we now transfer A from Susan to Marie. Susan loses 6 points of value and now has 67 points. Marie gains 5 points (remember, this is Marie's valuation, not Susan's) and now has 66 points.

You might think that we've gotten close enough—67 to 66—but to get things to be exactly equitable, we need to make another transfer. We transfer a small fractional ownership of item B from Susan to Marie. The idea is to find the fraction of item B that reduces Susan's value and increases Marie's value to the same amount. To save you the details, the fraction is about 1 percent. In other words, we transfer 1 percent of the ownership of item B from Susan to Marie, and this adjustment equalizes their valuations.

The final division of the three items is as follows:

> Susan: 99% of item B
>
> Marie: All of items A and C, and 1% of item B

Brams and Taylor show that the Adjusted Winner Procedure has some very nice properties. First, it is efficient, in the sense that any alternative allocation that improves one player's result will worsen the others. Second, it is equitable, as seen from the way that we carried out the transfer of items and fractional shares. Third, it is envy-free.

One problem with the Adjusted Winner Procedure is the same one that has plagued the previous two procedures: it is subject to strategic misrepresentation of the valuations. Because of this, Brams and Taylor sought to work out another procedure that would overcome, or at least reduce, the opportunity for players to falsify their reported valuations. This next procedure is called Proportional Allocation.

Proportional Allocation

Let's apply the Proportional Allocation Procedure to the same kind of fair division problem as with the Adjusted Winner method. There are two players (still Susan and Marie) and m items to divide up. Assume, to begin with, that the players honestly report their valuations for each item using points, so that the sum of their points over the m items is 100.

Proportional Allocation is simple and clear. For each item i, let S_i be Susan's valuation while M_i is Marie's valuation. The total valuation between the two players is $(S_i + M_i)$.

- Assign Susan her proportion $[S_i/(S_i + M_i)]$ of item i for all m items
- Assign Marie her proportion $[M_i/(S_i + M_i)]$ of item i, for all m items

Let's work out the Proportional Allocation for Susan and Marie using the same three items we just looked at.

	Susan's Valuations	Marie's Valuations	Total Valuation
A	6	5	11
B	67	34	101
C	27	61	88

- Susan is awarded $6/11$ of item A, Marie $5/11$
- Susan is awarded $67/101$ of item B, Marie $34/101$
- Susan is awarded $27/88$ of item C, Marie $61/88$

Now let's use an expected value calculation to check how many points each player has been awarded.

Susan receives $(6/11) * 6 + (67/101) * 67 + (27/88) * 27 = 56.0$ points

Marie receives $(5/11) * 5 + (34/101) * 34 + (61/88) * 61 = 56.0$ points

Notice that the 56 points that each player accrues here is less than the 66+ that they obtained via the Adjusted Winner Procedure. This lack of efficiency is a concern with Proportional Allocation.

However, the fact that both players come out with the exact number of points is no accident. Proportional Allocation always results in an equitable allocation, and on top of that, the allocation is always envy-free. Of course, even more than with the Adjusted Winner Procedure, Proportional Allocation depends upon the inherent divisibility of the items. In some applications, this simply can't be done.

Proportional Allocation was intended to reduce the temptation for the players to falsify their point distributions over the items to be divided. Brams and Taylor provide an argument to show, for a wide range of players' valuations over the different items to be divided, that truthful reporting in the Proportional Allocation Procedure gets close to the best strategy to carry out.

The Least You Need to Know

- Fair division methods treat items that are either inherently divisible or can't be broken up at all.
- A number of fair division methods for a single item have one player cutting up the item and the other players choosing pieces.
- Fair division methods focus on allocations being proportional, envy-free, efficient, and equitable.
- In the Knaster-Steinhaus Procedure, players assign monetary values to the items and may make payments that compensate for unbalanced allocations.
- Adjusted Winner and Proportional Allocation avoid some of the drawbacks of the Knaster-Steinhaus approach.

Cooperative Games

Chapter 12

In This Chapter

- Understanding cooperation and synergy
- Finding a formula for fairness
- Sharing costs and profits
- Keeping everyone together

Cost- and profit-sharing problems go far beyond the "I cut, you choose" method and the other fair division situations and techniques we considered in Chapters 10 and 11. In those problems, the cake (and the estate, for that matter) was a given. But now we realize that sometimes we have to work together to prepare that metaphorical confection. We each put in different ingredients and expertise, and this influences what we expect to receive when the handiwork is complete.

Pay Your Fair Share

It's 7:10 P.M. and you've just left your office. You're meeting friends uptown for dinner at 7:30 and you'll have to grab a taxi. In the elevator you're talking to your co-worker Jennifer, and it turns out that she is also planning to take a taxi uptown. Since you're going in the same direction, the two of you agree to ride together. When you get outside you see Dan, who works down the hall, hailing a cab. You all end up sharing a ride, and 20 minutes later you're the first one to hop out. The fare at that point (with a good tip) is $13. How much are you going to offer Jennifer and Dan?

The three of you shared the cab because doing so obviously lowers what each of you would end up paying. If you hadn't all run into each other, you would have taken three separate taxis, and now you've piled into just one. But how exactly will this translate into savings?

You believe that once you are out of the cab, Jennifer and Dan ought to shoulder the burden from that point on. So given that your outlay should apply only as far as your destination, what should you pay?

First, let's rule out an extravagant option: to produce a twenty and tell them to keep the change. Aside from being insulting, your $20 offer is not individually rational. As we saw in Chapter 10, a solution is individually rational if it leaves each player at least as well off as they would be fending for themselves. Since the fare at your destination is $13, your contribution should be less than $13 or there's no point in sharing the ride.

At the other end of the spectrum, suppose you pull out $1 and gingerly offer it while trying to look nonchalant. We need to tease out the economics from the emotions induced by this gambit. Jennifer and Dan will probably be taken aback. We will study situations like this later on, when we look at behavioral games. However, let's keep in mind that your measly dollar, if accepted, does in fact produce a small savings for them.

The reason that your miserly offer stirs up emotions is that it violates our sense of fairness. Game theorists who study these sorts of cost-sharing problems aim to devise solution procedures that are acceptable to all parties. Think for a moment and come up with a payment you would find reasonable, and we'll see what game theory has to say.

BET ON IT

Just because cooperative games sound warm and fuzzy doesn't mean there's no conflict. The players generate gains through efficiencies and synergies but they are still going to play fiercely to get what they can.

We're All on the Same Team

Game theorist Lloyd Shapley set out to solve a whole host of cost- and profit-sharing problems like our taxicab example. To see what he did, let's first back up a bit and think about what it means to share costs or profits in the first place.

In zero sum games, the players are practically in mortal combat with one another. What one player gains, the other loses. Nonzero sum games capture a much richer class of situations with conflicting strategies. But it's no longer wartime. The players now stand to benefit when they cooperate with each other.

Defining the strategies is a little harder now. With zero sum and nonzero sum games there were only a few strategies to choose from, but now there might be lots of alternatives. To capture the synergy and sharing aspects, and work out solutions to which players should pay what, game theorists developed models called *cooperative games*.

DEFINITION

Cooperative games are models of how different players benefit when they join forces.

To set up a cooperative game, you consider every possible grouping of the players and show what the cost or profit is when the rest of the players aren't in the picture.

The first thing you need to do is to establish who the players are. Right now this is easy: there's just you, Jennifer, and Dan. The next thing to do is to figure out what each of the individual players would have to pay if they acted alone. If you took your own cab, you would have paid $13. If Jennifer had taken her own cab, she would have paid more, say $18, since she was going farther. Finally, suppose Dan was going even farther and the fare at his destination was $27. These figures establish upper bounds on what each of the individuals will contribute.

The next step is to figure out what all the different combinations of players would have to pay. The whole point is that when various players join up, they lower their costs depending on how the different combinations change the circumstances. In doing this, you must consider what all the different subsets of players, which game theorists call coalitions, would pay if they "played" separately from the rest of the original group.

This step is called developing the *characteristic function* of the game. This essentially lists all the different possible groupings of players and what they would have to pay when acting alone. In our example setting up the characteristic function is pretty simple:

- You will pay $13 by yourself
- Jennifer pays $18 by herself
- Dan pays $27 by himself
- You and Jennifer together will pay $18
- You and Dan together will pay $27

- Jennifer and Dan together will pay $27
- You, Jennifer, and Dan will pay $27

> **DEFINITION**
>
> The **characteristic function** tells us what every single coalition is able to accomplish when acting alone.

There's one property the characteristic function will satisfy virtually all of the time: when different parties get together, they will do at least as well as when they were apart. For example, if you and Jennifer had never run into each other, you would have taken separate cabs and paid $13 + $18 = $31 altogether; but when you share a ride, your total is just $18.

The Shapley Value

To solve a cooperative game is to stipulate how much each player should pay in a cost situation, or receive in a profit situation. Any solution ought to satisfy two sensible conditions:

- **A solution must be individually rational,** as we have already seen. If a solution to a cooperative game is worse for a certain player than what she can obtain by herself, then why would she cooperate?

- **A solution must be feasible.** This means that the sum of the costs or profits given by the solution must equal the total bill or profit derived. This sense of feasibility is called "efficiency" by economists. An efficient solution is one that neither allocates more than what is available nor leaves any money on the table.

In 1953 Shapley devised a solution procedure for cooperative games in general. He was focused on coming up with a "fair" solution that is individually rational and efficient.

> **BET ON IT**
>
> Coming up with fair solutions is of paramount importance to game theorists. But rather than define "fair," they instead require that solutions obey certain reasonable conditions. It's up to us to decide whether the conditions are themselves acceptable.

Shapley approached the issue of fairness by coming up with the following three conditions that seem indisputable:

- **Players who add no value to the game should not benefit from it.** Shapley called such non-value-added players dummy players.

- **Changing the names of the players should not affect their allocations.** For example, alphabetical ordering or other labeling of the players should play no role in determining their shares. This condition is usually called symmetry or anonymity.

- **Additivity.** Suppose you had a fixed set of players who engaged in two different cooperative games. Now suppose the two games were combined together. Additivity means that each player's Shapley value in the combined game will equal the sum of their Shapley values from the individual games.

  Here's a nice way to illustrate additivity: suppose you go out to eat with the same friends twice. You could figure out, for each separate meal, how much each person has to pay, or you could combine the two checks into one overall bill and then figure out what everyone owes. Additivity says that the total cost for each person will be identical either way.

These conditions have to seem reasonable to you. Worthless players get nothing. No one gets special treatment because of his or her name. You might be tempted to make up some of your own statements that reflect what it means to be fair.

But the remarkable thing that Shapley did is to prove that there is only one formula that provides an efficient and individually rational allocation satisfying the three conditions.

I will demonstrate how to calculate the Shapley value soon. But before I show you the math, here's the idea. You know that cooperative games are meant to capture synergy that is often present when players combine their resources. For example, in the taxicab example, when you and Jennifer agree to share a cab, you both end up saving money. And when the two of you meet up with Dan, you save even more.

The Shapley value gives each player their average contribution when joining up with all possible groups. It turns out there is a nice way to measure a player's average contribution, or value added, when they join other coalitions. In a problem dealing with profits, this average value added is called a *marginal value*. In a cost environment, as in the taxicab situation, this average amount is called a *marginal cost*.

> **DEFINITION**
>
> A **marginal cost** is each player's average payment when joining up with all possible coalitions in a cooperative game. A **marginal value** is each player's average profit when joining up with all possible coalitions.

In the taxi example, the Shapley value has you paying $4.33, Jennifer paying $6.83, and Dan paying $15.83. In the following section I provide the details of the calculation. After that I show you that this particular solution has a nice, intuitive explanation.

Do the Math

Have you ever been a little bit slow getting your wallet or purse out, and while you were fumbling, someone else has paid the bill? Let's imagine this sort of scenario here. Suppose that the players ended up paying their contributions in a random order. One random ordering is Jennifer, Dan, then you; another ordering is you, Dan, Jennifer; and so on. For any particular ordering of players, the payment rule is to assign to them their marginal cost in that ordering.

For example, let's say you are first to pay your bill, then Dan, and finally Jennifer. You will pay $13 under this scheme because you are first to contribute toward the cost of the ride, and the fare when you got out was $13.

Dan shows up next in this particular ordering and contributes his marginal cost. Since the fare at Dan's destination is $27, he will have to pay the extra $14 by himself ($27 − $13 = $14). This is because Jennifer is chipping in last, and when she is ready to pay, her contribution at that point is zero since you and Dan have already paid the entire $27.

What we just did is to compute a cost allocation for just one ordering of the players. What if Jennifer paid first, followed by Dan, and then you? The cost allocation might be different. To deal with this, Shapley proposed that we find out every possible sequence of players, assume that the different orderings are equally likely (so no one player is favored to arrive first or last, for example), and then average up each player's marginal contribution over all of the different orderings.

Finding these average marginal costs sounds difficult, but it's actually not too bad unless you have a lot of players. With just three players, there are only six different orderings to examine. The following table makes it easy to keep track of everything. As a player makes a contribution relative to a particular ordering, we note down the contribution in that player's column in the table.

	Cost Shares		
Ordering of Players	You	Jennifer	Dan
You, J, D	13	5	9
You, D, J	13	0	14
J, You, D	0	18	9
J, D, You	0	18	9
D, You, J	0	0	27
D, J, You	0	0	27
Average Over All Rows:	26/6	41/6	95/6
=	4.33	6.83	15.83

Remember, the zeros in the table mean that at the point when a certain player was about to pay, their cost was already covered. So if Jennifer pays first, your fare is already included. And if Dan is the first to pay, he has to cough up the entire $27 and when that happens, both Jennifer's fare and your fare are taken care of.

The Analysis

Here's an appealing explanation of the cost shares. While you were in the taxi, three of you rode together. The fare was $13 when you got out. So you each pay ⅓ of $13, which is the $4.33 that Shapley indicates.

Then Jennifer and Dan ride on together until she gets out. The fare is an additional $5 ($18 − $13 = $5). Jennifer and Dan split that $5 segment. Therefore Jennifer pays $4.33 for the first segment plus $2.50 more, which totals $6.83.

Dan also owes $6.83 when Jennifer gets out, and from that point on he rides alone. He should pay the additional $9 from that point on, leaving him with a cost share of $15.83. Cool!

You might have noticed a certain "nested" property in the taxi example. Like Russian dolls, Dan's taxi ride includes Jennifer's shorter one, and Jennifer's segment includes yours. This property appears in real business applications, not just whimsical taxi examples. Suppose a light rail line is being considered that would link suburban communities A, B, and C to downtown, and suppose the cost to construct the rail line is a function of distance. In many instances the rail line will look like this (let D denote downtown):

With the rail line, the route from community B to downtown includes the link that serves community A, and so on. Since the same kind of nesting property applies, the Shapley value to provide cost contributions from the various communities to build the rail network will allocate its costs in exactly the same fashion as the taxicab computation.

Other Cost and Profit Examples

Most cost applications, however, do not exhibit this special rail or taxicab structure. In such cases the Shapley value can still be calculated in the same way, but the ultimate solution may not have the intuitively appealing property of the taxi solution.

An Application to Power Generation

Suppose three townships—A, B, and C—are planning to share the cost of a wind farm to be built nearby. The townships have different populations (both residential and commercial) and therefore different energy needs. In addition, the wind farm is not equidistant from the three townships.

Suppose A, if building its own wind farm, would need to spend $4 million; B, if building its own capacity, would have to spend $10 million; and C, $8 million. But because of the different fixed costs involved, if the townships teamed up, they would lower their costs.

In particular, a wind farm that would supply all three townships would cost just $19 million, instead of the $22 million if three were built separately. Suppose the power company also quoted you the costs for wind farms that would supply all the different pairs of townships. This information is summarized here, and yields the characteristic function of the game.

Price quotes from the utility company:

- A's cost is $4 million
- B's cost is $10 million
- C's cost is $8 million
- A and B together would pay $12 million
- A and C together would pay $11 million
- B and C pay $16 million between them

If the three townships were to fund a wind farm that would serve all three of them, it would cost $19 million. But the question is, given the cost data, how much should each township contribute? Even though this example does not exhibit the convenient "nested" property of the taxi or light rail applications, it can still be easily handled. The Shapley value is our weapon here, and the details of its calculation are in the following table.

Ordering of Players	Cost Shares		
	A	B	C
ABC	4	8	7
ACB	4	8	7
BAC	2	10	7
BCA	3	10	6
CAB	3	8	8
CBA	3	8	8
Average Over All Rows:	19/6	52/6	43/6
=	3.17	8.67	7.17

Calculating the Shapley value for the wind farm problem.

The Shapley cost allocation has A paying $3.17 million, B paying $8.67 million, and C paying $7.17 million. Notice how each of the three townships comes out better when they cooperate as opposed to what they'd have to pay when going it alone: A pays $3.17 million instead of the $4 million by itself; B pays $8.67 million instead of its stand-alone cost of $10 million; and C pays $7.17 million instead of $8 million going it alone.

Another thing to think about is this: suppose the township directors had never heard of game theory. Imagine how they would fight over the cost allocation! The townships might never even come to an agreement, and in that case perhaps the wind farm would never have materialized. The Shapley value acts as an impartial arbitrator that all parties can trust. They know that the resulting allocation is fair to all.

POISONED PAWN

The Shapley value is easy to do by hand with 3 or 4 players, but starts to get out of hand with more players than that. For more than a dozen players, it's only possible to compute the Shapley value using sophisticated approximations, if at all.

Sharing the Profit

Before we move on to another solution concept, let's briefly see how the Shapley value can be applied to profit-sharing situations, using an example of a simple market.

Suppose you own a horse, which is worthless to you unless you can sell it. You bring the horse to a fair and there are two people who would like to buy it. Buyer 1 values the horse at $200. In other words, Buyer 1 is willing to pay up to $200 for it. Meanwhile, Buyer 2 values the horse at $250.

To see how to model the horse market as a cooperative game, suppose you sold the horse to Buyer 1 for $150. You make $150 while Buyer 1 gets a $200 horse for $150, a gain of $50. Together you have gained $200 from this transaction. In fact, for any transaction amount between $0 and $200, you and Buyer 1 together will generate a profit of $200. Similarly, selling the horse to Buyer 2 generates a profit of $250 between you.

Also realize that no profit is garnered unless you sell the horse to one of the potential buyers. Therefore no single player generates any profit, and neither does the coalition consisting of the two buyers.

Let's summarize this information in characteristic function form:

- Your profit acting alone is $0
- Buyer 1's profit acting alone is $0
- Buyer 2's profit acting alone is $0
- Buyers 1 and 2 generate $0 profit
- You and Buyer 1 generate $200
- You and Buyer 2 generate $250
- All three players together generate $250

To generate the maximum possible profit of $250, Buyer 2 must end up with the horse. Let's see what the Shapley value of the game is, and we'll be able to discover some underlying economic meaning.

The Shapley value hands you a $158.33 profit, while Buyer 1 receives $33.33 and Buyer 2, $58.33. As we can see, the Shapley value splits up the $250 total, but what happened to the horse?

	Profit Shares		
Ordering of Players	You	1	2
You, 1, 2	0	200	50
You, 2, 1	0	0	250
1, You, 2	200	0	50
1, 2, You	250	0	0
2, You, 1	250	0	0
2, 1, You	250	0	0
Average Over All Rows:	950/6	200/6	350/6
=	158.33	33.33	58.33

Calculating the Shapley value for the Horse Market game.

The most direct way to generate $250 worth of profit is for you to sell the horse to Buyer 2. But why should Buyer 1 profit? Buyer 1 is a losing bidder ... what economic role does he play?

The numerical analysis shows us that one of the market scenarios occurs when Buyer 1 buys the horse from you, and thereafter sells the horse to Buyer 2. In that case Buyer 1 has acted as a "middleman" and thus extracts some value.

Going beyond that, if you are Machiavellian (and this *is* a game theory book), you'd realize that Buyer 1 could have been your confederate. As your partner, his role in the market would have been as a shill: to jack up the price that Buyer 2 ultimately has to pay. While the Shapley value does not explicitly consider morality or intention—it is simply a product of the numerical input—it does assess an economic value to Player 1 in this case: an agency fee, if you like, of $33.33.

Trying to Satisfy Everyone

Nifty as it is, the Shapley value is not the only "fair" way to try to solve cooperative games. In the same year that Shapley published his work, Donald B. Gillies came up with an idea called the *core* of a cooperative game.

Remember that individual rationality made sense, and so did efficiency. Gillies asked for one more condition to be satisfied: coalitional, or group, rationality. Gillies wants all coalitions to do at least as well if they play the game along with everyone else as they would on their own.

> **NO BLUFFING**
>
> An extraordinary amount of pioneering work in game theory came out of Princeton University in the 1940s and 1950s. John von Neumann was there at the Institute of Advanced Study. He and Oskar Morgenstern, an economics professor, wrote the first textbook on game theory. John Nash, Lloyd Shapley, and Donald Gillies were graduate students, and much of their work was supervised by professors Harold W. Kuhn and Albert W. Tucker, each of whom made a number of important contributions to game theory and other fields.

In other words, any reasonable solution to the game ought to dish out a cost or profit share to every coalition that is at least as good as what that coalition can achieve on its own (that is, without cooperating with others). Any such solution that can't be bettered by any coalition acting on its own is called a core solution.

If we can come up with a core solution, it not only keeps the individual players happy, but keeps every possible coalition happy, too. No coalition will have any incentive to defect from the group if presented with a core solution, because they can't do any better by themselves. A core solution would act as the bond that keeps the players from defecting.

The core of a game sounds like an unshakably good idea. But there are two basic facts you must remember when looking for the core of a game. One is that the Shapley value, unfortunately, might not be a core solution. But even worse, sometimes there might be no allocations at all that satisfy the core conditions. In other words, the core might be empty.

In the horse market example, the Shapley value awarded you $158.33 and awarded Buyer 2 $58.33. So between you, the Shapley value allocates $216.66.

But remember that you and Buyer 2 could have transacted the horse without Buyer 1 and gained $250. Because of this, the two of you together would be unhappy with the Shapley allocation totaling just $216.66. In other words, the Shapley value is not a core solution. This mismatch is a concern because a solution that violates the core is subject to protest by one or more coalitions.

> **POISONED PAWN**
>
> The Shapley value often violates the core requirements. Other solution concepts have been proposed, but they all have some deficiency. This does not diminish these solutions; it simply means there's no one-size-fits-all approach.

It's important to note that sometimes there is no solution that satisfies the core conditions. In such cases you would probably want to resort to the Shapley value and realize that there will always be some coalitions that will be frustrated.

For example, suppose you, Jennifer, and Dan are walking around together and find $120 on the ground. This happened to me years ago in San Francisco; my two friends and I used the money to buy dinner. But let's say that you and your friends are democratically inclined and decide to split the money up according to a kind of majority rule system.

What you do here is, you decide that any majority of the three of you "deserves" to keep all $120. Let's spell out the consequences of this seemingly democratic rule:

- You and Jennifer should get the $120 between you.
- You and Dan ought to get all $120 between you.
- Jennifer and Dan also deserve the full $120.

Is there any possible way to satisfy all of the coalitions here? You can bang your head against the wall trying, but the unfortunate truth is that no such allocation can be found.

On the other hand, if you simply divided up the $120 equally (which happens to be the Shapley value), neither you, Jennifer, or Dan should have any worries about getting a raw deal.

To sum up, a great many cost- and profit-sharing situations can be modeled as co-operative games. The Shapley value provides a neat way to solve these games, but sometimes the Shapley value violates another reasonable solution, the core. And asking for a solution to obey the core conditions is sensible, too, except sometimes there's just no way to do it.

The Least You Need to Know

- Cooperative games model situations in which more than two players join forces to increase profits or lower costs.
- Solutions to cooperative games show whether individual players are better off when they team up with others.

- There exist at least two solution approaches to cooperative games—the Shapley value and the core—that are based on intuitively appealing assumptions.
- Sometimes the Shapley value conflicts with the core, and sometimes there is no core solution at all.

Group Decision Games

Chapter 13

In This Chapter

- Identifying the pitfalls of different voting systems
- Viewing voting as a cooperative game
- Making sense of veto power and dictatorship
- Measuring voter influence
- Considering one promising method
- Understanding impossibility theorems

Voting is perhaps the ultimate group decision. Whether voting on committees, in the boardroom, or in state elections, people across the world embrace voting as the key feature of democratic government. The act of voting is a powerful symbol, especially in countries transitioning from autocratic to democratic governance. Yet as much as having a vote gives citizens power, the fact that an election may involve millions of people often seems to render each individual vote meaningless.

This chapter investigates voting processes and outcomes from a game-theoretic standpoint. By analyzing features like coalition building, veto power, proportional representation, and other familiar democratic concepts from a strategic point of view, you will uncover some surprises.

Voting as a Group Decision

Imagine a group of people trying to make a decision. It's easy to see how personality, temperament, and social and communication networks would play a role as the group moves toward a decision.

After discussion and debate, the time will come to make a choice. In extreme cases, one person will take over the proceedings and decide what to do. We normally call such a person a dictator, a notion most people raised in the West find to be abhorrent. The more usual practice is for each person to have a say in the outcome, and the normal way to elicit this opinion is through a voting mechanism.

Veto Power

Before exploring different voting systems, let's first consider a few situations where one player essentially holds all the cards despite the group nature of the process.

In Chapter 12 you learned how to model economic situations with multiple players by setting up a characteristic function that evaluated what each coalition of players was able to attain. This cooperative game approach is useful in studying voting as well. Let's consider voting situations that are decided by majority rule. This means that any coalition, or subset, of players that has more than half of the total number of votes is a winning coalition. A simple way to represent this using the characteristic function is to assign a value of one to all winning coalitions and a value of zero to all the others.

Now let's examine a curious game that will lead to some insight on players' power in voting situations. Consider a school that has collected 100 pairs of gloves in a large box to be donated to a local charity. One afternoon, a mean-spirited boy, Kid X, breaks open the box. He steals all 100 right-hand gloves and scatters all the left-hand gloves around the school. On his way home, though, he accidentally drops a few gloves in the river. At home he counts 96 gloves.

The next day, the missing gloves are noticed. The school decides to announce a $1 reward for each left-right pair of gloves that is returned. That afternoon, 100 different kids have retrieved all 100 left-hand gloves. Kid X shows up with 96 right-hand gloves and no one thinks this is at all suspicious. The other children then line up to bargain with Kid X individually.

Now imagine each single kid haggling with Kid X. These kids will think that they should share the $1 evenly with him. But Kid X tells each one he'll only offer 10¢. Consider an indignant child who believes that only an offer of 50¢ is fair. They threaten to walk away, and Kid X says, "I don't need you." The thing is, Kid X is right. No individual child has any clout with Kid X, because if that child walks, there are 99 other kids and only 96 right-hand gloves anyway.

BET ON IT

The interesting feature is that if Kid X had held on to all 100 gloves, he would not have been in the same strong bargaining position. It is precisely his scarcity of gloves that strengthens his position, rendering any other individual superfluous.

In fact, in this game Kid X holds all of the power, unless a large group of the others bands together. In a sense, even the 10¢ he is offering is generous, and no individual can threaten or persuade Kid X to raise his offer.

Now let's look at a much smaller version of this game and analyze it using cooperative game theory. Suppose there are three players, called R, L1, and L2. Player R has a right-hand glove and players L1 and L2 hold one left-hand glove apiece. Any L-R match wins $1.

Let's write down the characteristic function for this game:

- Player R gets $0 if acting alone
- Player L1 also gets $0 if alone
- Player L2 gets 0 when alone as well
- Players L1 and L2 get $0 (two left gloves)
- Players R and L1 get $1 when paired up
- Players R and L2 also get $1 when paired
- R, L1, and L2 also generate a value of $1

This type of game is called a simple game: all coalitions have characteristic function value equal to 1 or 0.

Notice that the only coalitions with a value of 1 are those that include Player R. Recall that the core of a game is an allocation that is individually rational and group rational. There is a total of only $1 to allocate. But both the coalitions (R, L1) and (R, L2) "deserve" this $1. If L1 gets anything, then R and L2 together end up with less than the $1 they should be awarded. Similarly, if L2 gets anything, then R and L1 together get less than their $1.

The only core allocation, then, is for Player R to receive the entire dollar while players L1 and L2 each get nothing. What are the implications for voting situations?

In a voting context, coalitions with value one are winning coalitions, and those with value zero are losing coalitions. What is special about Player R is that no coalition can win without him. Player R, then, has a special role: R is called a *veto player*. Since no other coalition can get anywhere without R, R has a great deal of power. While veto players are not quite dictators, they are indispensable, and certainly they wield disproportionate power.

> **DEFINITION**
>
> A player in a simple game is called a **veto player** if that player is a member of all winning coalitions.

By the way, whatever happened to majority rule? The lesson here is that even though players L1 and L2 form a majority of the players, the actual power they exercise in the game is limited because of the way the game is defined. As you will see, it is often the case that certain players or coalitions have much more, or much less, power than it initially seems.

While the cooperative game structure proved useful in analyzing the veto game example, it is not the only valuable approach to voting games. As I pointed out in Chapter 9, some voting rules assign points to a ranking of alternatives, while others are based on a fundamental principle like proportionality. I walk you through a variety of voting games in the following sections.

Disappointments and Surprises

Modern voting theory was developed in eighteenth century France by two figures of the Enlightenment, Jean-Charles de Borda and the Marquis de Condorcet. Their analysis and methods are very much relevant today.

Pairwise Comparisons

Condorcet wanted to use comparisons between pairs of alternatives (often referred to as "pairwise comparisons") as the basis of a voting procedure. He proposed that when voters choose among a set of alternatives, the winner ought to be the choice that would gain a simple majority of votes in a two-candidate runoff against every other alternative. Intuitively, the Condorcet winner—the candidate that would beat every other one it was matched against—sounds like a solid pick.

Although there are many voting situations in which the Condorcet winner is clearly the best choice, there are problems in trying to rely on this notion. For one thing, in some voting instances there is no Condorcet winner. But what follows is a bigger pitfall.

Suppose there is a population of 300 voters considering three candidates A, B, and C. They each list their preferences on a ballot, and when all the ballots are counted, we find the following:

- 100 voters prefer A, then B, and C last
- 100 voters prefer B, then C, with A last
- 100 voters prefer C, then A, then B

First of all, notice that there is a three-way tie among A, B, and C. But that's not the strange thing. Two hundred out of the three hundred voters—a clear majority—prefer A to B. But then 200 out of 300 prefer B to C, and 200 out of 300 prefer C to A. Notice how these preferences cycle around. I illustrate that these preferences are *intransitive* in the following diagram, where an arrow signifies preference. The type of cycle illustrated in the diagram is sometimes called a Condorcet cycle.

A Condorcet cycle illustrating intransitivity.

 DEFINITION

For three alternatives A, B, and C, if A is preferred to B and B is preferred to C, we would expect A to be preferred to C. If instead C is preferred to A, we say that the preferences are **intransitive.**

Individual Rankings

Condorcet's contemporary and rival Borda wanted to derive a voting procedure based on the positions of the candidates in individual's rankings. His idea was that an alternative that was ranked first by a voter would receive more points than the one ranked second, and so on. Borda proposed aggregating the points gained by the candidates among all of the voters, with the winner simply being the candidate with the most points.

It turns out that Borda methods are no more impervious to paradoxical results than the Condorcet methods are. The so-called "winner-turns-loser paradox" is a particularly surprising twist. Suppose we have four candidates, and a bunch of voters, each of whom ranks the candidates from first (giving them four points) to last (one point). We total up the points among all the voters and rank the candidates from first to last, with a clear winner.

Now the fun starts. Suppose the candidate who finished last is eliminated from the election just as it is concluded, and the election commissioners decide to recount the votes among the first three finishers. The votes are not recast. Rather, the exact same ballots are used but now the candidate ranked highest gets three points (instead of four), and so on. It is possible that the candidate who placed first when there were four of them will no longer be the winner when the last place finisher is removed from the ballots!

> **BET ON IT**
>
> Borda point-count tallies are used in a number of voting situations today. One example is the ranking of college basketball teams in the United States by a group of sportswriters. Each sportswriter writes down, in order, their list of the top 25 teams. The team ranked number 1 gets 25 points, the second ranked team gets 24 points, and so on. The points are added up among the different sportswriters and the team with the highest point total is ranked number one, the team with the second highest total is ranked number two, and so on, down the line.

This result seems to contradict the "independence of irrelevant alternatives" condition we saw in the context of developing a bargaining solution (see Chapter 10). On the face of it, deleting the worst candidate should not affect the ones at the top. But in this voting context, the candidate that ranked last is not necessarily irrelevant. When this candidate is deleted, on many ballots other candidates might move up a notch, and this realignment can actually knock the original winner from the throne.

Finally, note that both the Condorcet and the Borda approaches are subject to strategic manipulation. I mentioned in Chapter 9 how voters in a presidential election might not vote for their top choice if they felt the candidate had no chance of winning. Instead they might shift their allegiance to their second-best choice. It turns out that Condorcet methods and Borda systems are both vulnerable to deliberate voter misrepresentation of their true preferences.

Problems with Proportionality

In Chapter 9 I raised the question of whether proportionality could be preserved in a voting system. When the state of Pennsylvania, with 4 percent of the U.S. population, would be delegated 19.6 electoral votes if we used fractions, we need to know whether to round Pennsylvania up to 20 or down to 19. But many other states would also end up with fractional numbers of electoral votes. Given that the total must remain at 538, it becomes a tricky—often impossible—task to round all of them up or down in a consistent way and stay at the necessary 538.

This problem of rounding sounds unfortunate, but not disastrously so. However, there are other problems with this kind of proportional apportionment, and these pitfalls are both surprising and troubling.

As it grew, the United States kept adding states (until 1959). Along with the growing number of states, the United States kept increasing the size of the House of Representatives. After the 1880 census a peculiar problem was discovered: when going from 299 representatives to 300, the state of Alabama's apportionment actually went down from 8 to 7!

It turns out the source of the problem was the way Alexander Hamilton had solved the problem of fractional apportionments. His method, in fact, seems quite reasonable: first compute every state's proportion of representatives, and give each state its integer part of that proportion. Now add up the allocated seats. If there are any leftovers, give these out in order to the states with the highest fractional parts. To see how this works, suppose in a country of just 3 states and 100 seats, the proportions came out as follows:

- State A   19.5
- State B   48.1
- State C   32.4

Hamilton's rule first gives State A 19, State B 48, and State C 32 seats. Then there is one leftover seat. We give the leftover to the state with the highest fractional part. In this case State A gets the extra seat. The problem that befell Alabama was that when the number of representatives grew, another state gained a larger fraction than Alabama did, and Alabama lost the extra seat.

Since Hamilton's apportionment rule, though reasonable, can run afoul of problems like the Alabama paradox, how about using a different apportionment rule? A number of luminaries, including Thomas Jefferson and Daniel Webster, proposed alternative apportionment schemes. But other methods can suffer from a different problem: the so-called *population paradox*, a situation in which it is possible that State A can lose population, State B can gain population, and yet State A gains a seat at the expense of State B. It turns out, unfortunately, that no one has ever devised a rule that avoids all such problems.

Proportionality and Power

Earlier in this chapter you learned how a veto player can hold seemingly disproportionate power. Now we will probe more deeply into what voting power really means. In many voting situations, individual voters or blocs might hold more than one vote. For example, at a shareholder meeting, some voters hold more shares than others and thus would seem to be more influential. Certain states in the U.S. presidential election carry more electoral votes than others, and similarly appear to be more important.

We will look at two different ways to determine whether the true power these entities possess has anything to do with their proportionate shares. In these cases, each player has a certain assigned weight—these weights represent, for example, that player's number of shares, or their relative proportion of the total population.

Remember, though, that the notion of proportionality not only strikes most people as "fair" but is emphasized in the U.S. Constitution under the principle of one person, one vote. We will explore whether ordinary voting systems actually adhere to this principle.

In the following situations, voters do not assign points or indicate preferences over a slate of several alternatives. Instead, they vote only yes or no. This binary choice can represent whether voters are voting either for or against a particular motion or measure to be passed, or are voting for one or the other of two candidates.

In Chapter 9 I introduced an example in which a company had four shareholders. Let's reproduce the situation:

- Player 1 owns 100 shares
- Player 2 owns 100 shares

- Player 3 owns 80 shares
- Player 4 owns 50 shares

Since the total number of shares is 330, the board of shareholders has enacted a rule that for any motion to pass, it needs at least 170 votes.

Let's now quantify each player's (i.e., shareholder's) amount of relative strength in the game with a measure called a *power index*. A naïve, but easily understandable power index is found by computing each shareholder's proportion of the total number of shares. According to this proportional measure:

- Player 1 holds $100/330$ = 30.3 percent of the power
- Player 2 also holds $100/330$ = 30.3 percent of the power
- Player 3 holds $80/330$ = 24.2 percent of the power
- Player 4 holds $50/330$ = 15.2 percent of the power

With a proportional measure, Player 4, for example, would expect to have 15 percent of the power when the board convenes to make decisions. But as you saw in Chapter 9, Player 4 in fact has no real power or influence at all.

DEFINITION

In a voting game, a **power index** measures each player's relative strength, influence, or control.

A **swing vote** occurs when a coalition loses without that player but wins with them.

To see this, look at all of the winning coalitions—these are the groups of shareholders that total 170 or more shares:

Coalition {1,2} is a winning coalition; so is {1,3} and also {2,3}. In addition, {1,2,3}, {1,2,4}, {1,3,4}, {2,3,4}, and {1,2,3,4} are winning coalitions. But notice there is no situation in which Player 4 has a *swing vote*. Look at coalition {1,2,4}, for example. Players 1 and 2 are a winning coalition without Player 4. In other words, Player 4 is useless to players 1 and 2. The same thing turns out to be true for coalitions {1,3,4}, {2,3,4}, and also {1,2,3,4}.

Player 4 is never needed to create a majority. In other words, she never swings the vote. The consequence of this is that Player 4 has no power at all to pass, or veto, any motion made by the board. It is a mistake to believe that Player 4 has 15 percent of the power on this board; in fact, any reasonable power index ought to assign Player 4 0 percent.

In 1965 John F. Banzhaf III sought to prove that the Nassau County, New York, board had a similarly unfair voting system. Here's the original data that prompted a lawsuit. Nassau County had six different municipalities represented. These six players in the game were assigned votes in proportion to their relative populations.

Player	Label	Votes
Hempstead 1	A	9
Hempstead 2	B	9
N. Hempstead	C	7
Oyster Bay	D	3
Glen Cove	E	1
Long Beach	F	1

There are a total of 30 votes. The rule was that any majority of at least 16 votes would be required for passage. What Banzhaf did is to develop a power index along the lines of what I began for the four players in the preceding example. Here are the steps:

1. Identify all possible winning coalitions.
2. In each winning coalition, identify all players that provide a swing vote.
3. Count up all swing votes tallied in step 2.
4. For each player, divide its total number of swing votes by the overall sum. This ratio is the Banzhaf Power Index.

In the following section I work out the details of calculating the Banzhaf Power Index for the six-town example. Notice how all the small townships equally have no power at all, while the three bigger ones, despite unequal size, all have equal $1/3$ weight.

Do the Math

Out of the 64 possible coalitions, there are 32 winning ones. These are listed on the following page. In any winning coalition some of the players may have swing votes while others do not. Every player that has a swing vote in a certain coalition is in bold.

AB, **AC**, **BC**, ABC, **ABD**, **ABE**, **ABF**, **ACD**, **ACE**, **ACF**, **BCD**, **BCE**, **BCF**, ABCD, ABCE, ABCF, **ABDE**, **ABDF**, **ABEF**, **ACDE**, **ACDF**, **ACEF**, **BCDE**, **BCDF**, **BCEF**, ABCDE, ABCDF, ABCEF, **ABDEF**, **ACDEF**, **BCDEF**, ABCDEF.

There are a total of 48 player labels in bold, and these are all of the swing votes that occur. Notice that many coalitions do not have any swing votes. For example, the coalition ABC does not have a swing vote because none of the three players is indispensable.

Player A (Hempstead 1) appears in bold 16 times. Therefore A can swing the election in 16 out of the 48 winning coalitions. So A has a Banzhaf Power Index of $^{16}/_{48} = ^{1}/_{3}$.

Player B is symmetric to A and also has a Banzhaf Power Index of $^{1}/_{3}$.

Player C, despite its smaller proportional representation, also has a Banzhaf Power Index of $^{1}/_{3}$ (count 'em!). And finally, none of the three smaller players is ever a swing vote, so all have power equal to zero.

Other Power Indices

The Banzhaf Power Index is not the only such measure. Lloyd Shapley and Martin Shubik proposed a different power index that predated Banzhaf's. The Shapley-Shubik Power Index also counts swing votes, but the swing votes are counted by using the Shapley value for cooperative games that we saw in Chapter 12. In other words, the Shapley-Shubik Power Index is just the Shapley value in the special case of simple games.

> **NO BLUFFING**
>
> During election campaigns in the United States, political pundits talk incessantly about "swing states," so it is not surprising that there is a history of analytical methods to capture this notion. The first known publication on what we are calling a power index was by L. S. Penrose (a British geneticist) in 1946. Shapley and Shubik developed their index in 1954. Banzhaf, unaware of the previous work, essentially rediscovered the Penrose method in 1965, and ditto for James S. Coleman who reiterated the same technique in 1971.

Since we worked out Shapley value examples in Chapter 12, we will not do more of them here. In some cases, like the two preceding examples, the Shapley-Shubik Power Index will match the Banzhaf results exactly, though in many cases it will not. Generally, the results from the two methods will not diverge much.

The difference is found in the way that the Shapley value generates the orderings of players. While the Banzhaf index looks at a coalition once to count up the swing votes, the Shapley value also considers the different ways that the coalition could have formed, thus assigning a different weight to a particular swing vote. That aside, for most experts there isn't much advantage using one method over the other.

The U.S. Electoral System

Now let's consider one of the biggest games of all—the U.S. presidential election—and determine if the rules of the election truly follow the one person, one vote principle set down in the U.S. Constitution.

We already saw, in Nassau County, how residents of townships D, E, and F had no power at all. Therefore, the principle of one person, one vote certainly did not hold in Nassau County politics.

What about in the U.S. presidential election? The 538 electoral votes are distributed among the states in a roughly proportional way. Each state gets 2 votes to begin with and then receives extra votes according to its proportion of the population (subject to Webster's apportionment rule). Washington, DC, receives 3 votes. A candidate needs at least 270 electoral votes to claim victory.

Now let's think about the chances that a state will be a swing state. A swing state is the one that puts the winning candidate over the top. In the 2000 election Florida was the swing state that decided the election in favor of George W. Bush.

Without doing any analysis, it should be clear that there are vastly more opportunities for large states to be swing states than for small states. A state like Florida with 25 electoral votes can swing the election for a candidate if that candidate has already accumulated anywhere from 245 to 269 electoral votes; a small state like Delaware with 3 electoral votes can only swing the outcome when the candidate has 267, 268, or 269 votes.

So it is clear that Florida, for example, will have a much higher power index as a state than does Delaware. No surprise there. But then, Florida has a population about 20 times that of Delaware. To compare individual voter power, we have to divide each state's power index by its population. What we want to find out is, do individual voters in Florida or Delaware (or the rest of the United States) have unequal influence on the outcome of the election?

Game theorist Guillermo Owen has answered this question. Although any power index will change with each census, Owen showed, using both the Banzhaf index and the Shapley-Shubik index, that voters in large states can have as much as three times the influence on the election outcome as voters in small states. This certainly seems to violate the one person, one vote principle.

While this result seems discouraging, there is an extraordinarily simple solution. If the United States abandoned the electoral college and simply relied on counting the popular vote, it would solve the problem of voters in different states having unequal power. Doing away with the electoral college would also solve the even bigger problem of allowing a candidate who loses the popular vote to win the electoral vote and the election with it.

NO BLUFFING

Since 1824, three different presidential candidates have lost the popular vote but won the election: Rutherford B. Hayes in 1876, Benjamin Harrison in 1888, and George W. Bush in 2000.

Approval Voting

Thus far we have looked at different ways to represent voter preferences: using pairwise comparisons among alternatives and ranking and assigning points. We have seen what it takes to be a powerful veto player, and we have also seen that proportional shares do not translate into proportional power. What is most interesting, but also discouraging, is that we have run into a number of paradoxes and pitfalls. Voters can manipulate any kind of system we set up. Preference rankings can go intransitive. Districts can have too little power or too much. Isn't there some positive news here?

The answer is yes. Instead of ranking all candidates and assigning points, or forcing yourself to agonize over a decision between two candidates that you like almost equally, why not just vote for all of the candidates that you approve of? This way of voting is called approval voting, and is generally attributed to Steven Brams, Peter Fishburn, and Robert Weber. The idea is that the candidate with the most votes wins, and the interpretation is that this candidate is the most acceptable choice to the largest number of voters. While approval voting is not a perfect system, it avoids a lot of the problems that we have seen.

The advantage of approval voting occurs when there are three or more alternatives. If you favor a candidate who is running a distant third, under approval voting you would not have to worry about being pragmatic and voting instead for one of the more mainstream choices. You can have your cake and eat it too!

So if you are environmentally focused but are afraid that your vote would be wasted on the green party candidate, you can vote green and go for a mainstream choice, too. Similarly, if you are a fiscal conservative and favor a libertarian candidate, you can still support him as well as any acceptable choice from the two main parties. Because of this, approval voting would significantly cut down on (although not eliminate) strategic manipulation. In addition, approval voting would prevent parties from getting torn between two of their own candidates. Moreover, at least in the United States, approval voting would open up elections to more candidates. (In fact, one criticism of approval voting is that it might destroy the two-party system.) Brams and Fishburn point out some other advantages: approval voting gives voters more options; approval voting might increase voter turnout; it would help elect the most favored candidate. Finally, it is easy to understand and simple to implement.

Again, approval voting is not perfect. It is subject to strategic misrepresentation, although not nearly as much as other voting methods are. Perhaps the biggest drawback is that, among a voter's set of acceptable candidates, there is no way to express that one candidate is more preferred to another.

The Search for a Perfect Method

Here it is in a nutshell: there is no perfect method, unless dictatorship is your cup of tea. In a celebrated work, Nobel Laureate Kenneth Arrow proved an "impossibility theorem" with respect to finding an ideal method. His idea was to study voting in an axiomatic fashion to see whether any system might just fit some indisputable conditions. Here are Arrow's axioms that any reasonable voting process ought to obey. We assume there are at least three alternatives present:

- Each voter is able to rank all of the candidates (ties are allowed).
- Each voter's ranking is transitive.
- No voter is a dictator (i.e., no single voter's preference can solely determine the outcome).

- If everyone prefers A to B, then A must end up ranked higher than B (this is called unanimity).
- Independence of irrelevant alternatives (preferring A to B is independent of the presence or absence of other choices).

Arrow's theorem states that no voting method can exist that satisfies all of these conditions. Put differently, Arrow's theorem says that any voting system that is transitive, and obeys unanimity and independence of irrelevant alternatives must be a dictatorship!

Arrow's theorem tells us what people suspected for a very long time anyway. Perhaps Winston Churchill put it best when he said that democracy is the worst form of government except for all the others. But before we leave this topic, let's nail down one more intriguing issue: can a voting system be designed that can't be manipulated by the voters? In other words, is it possible to set up a voting procedure that will induce all voters to act according to their true preferences?

Alas, in the 1970s, two different researchers, Allan Gibbard and Mark Satterthwaite, showed that any nondictatorial voting scheme is subject to strategic manipulation by the voters. So while we have been repeatedly frustrated by the fact that each voting method we look at can be manipulated, at least the Gibbard-Satterthwaite theorem informs us that we need not bang our heads against the wall seeking this particular holy grail.

The Least You Need to Know

- Voting systems that use pairwise comparisons or point ranking systems are sensible but not foolproof.
- Proportional representation may not yield proportional power.
- Approval voting avoids some problems of other methods.
- No voting system is perfect.
- Any voting system can be manipulated.

Part 4

Individual Values vs. the Group

In this part the clash between what individuals want and what's good for the group is starker than before. For instance, when players share a common resource, if they each use that resource according to their own best interest, the resource will eventually be depleted. However, if the players figure out a way to manage that resource for everyone involved, they can conserve the resource while still benefiting from it. Yet a different temptation arises when private information entices people to cheat others.

Game theory provides weapons for dealing with problems like these. The overall strategy is to lift the fog of incomplete information. One way to do this is through the use of auctions. Auctions are powerful tools that tease out the private information that players like to keep hidden.

The most surprising weapon in your arsenal, though, is something called mechanism design. The idea here is to first identify a desirable outcome and then to reverse-engineer the rules of the game so the players will have incentive to behave themselves, so to speak. As you'll see, mechanism design provides some clever ways to circumvent the incentive problems you've learned about, and to implement solutions for the greater good.

Individual Gain vs. Group Benefit

Chapter **14**

In This Chapter

- Harnessing pooled resources
- Sacrificing oneself for the common good
- Making sense of the Free Rider problem
- Seeing what happens in the real world

One theme that runs through much of game theory is the clash between individual desire and what is best for the group. Often what one person wants will undermine the outcomes for everyone else. In this chapter and the two that follow, I focus on the interaction between individual rational play and group optimal outcomes. The Stag Hunt game in Chapter 5, in which an individual is tempted to chase a rabbit for himself as opposed to teaming up to hunt deer, is a good introduction to this theme.

This chapter explores the Tragedy of the Commons as a model of friction between individual and group strategies. It also looks at other important game models that speak to economic, moral, and ethical dilemmas that we face in routine social interaction. The final section of this chapter draws examples from the real world to see whether people's actual behavior in the real world is in line with what theory predicts.

The Tragedy of the Commons

When ecologist Garrett Hardin described the Tragedy of the Commons in 1968, it was already a well-known phenomenon. Hardin used the concept as a way to describe overuse of publicly held resources, such as land, air, or water. Consider, for instance, what happens in a rural community to land that is held in common and open to all. If ranchers are allowed to graze their cattle on the land with no management practices in place, the land will eventually be totally depleted. The Tragedy of the Commons can also be used to describe other practices, such as overfishing and deforestation, which are difficult to control.

Hardin's two solutions to the Tragedy of the Commons span the economic/political spectrum. One solution is to privatize the land by selling it off in plots. The land will now be privately owned, and sensible owners will take care not to squander their own resources.

At the other extreme, the government could take over and manage the land. Just as we would expect individual landowners to do right by their property, in the case of government regulation we would trust that the resources would not get depleted as they did when the commons was a free-for-all.

It is useful to conceptualize the Tragedy of the Commons as a multiperson game. But the situation needs to be a bit better defined. We are not talking about mining or oil drilling, for instance. In a mine or oil field the total resource available is finite, and once such a supply is depleted, it's gone forever. Instead, let's focus on resources like fish and timber that are renewable, at least, if managed wisely.

The Commons Game with a Low Threshold

In many situations in which there is a common resource players typically follow one of two basic strategies. The first is, in essence, a greedy strategy—to maximize what they can extract. The second strategy is to go easy—to harvest the resource with restraint. Let's call these strategies G for greedy harvesting, and M for modest harvesting. In the following game, assume that just one player, by choosing a greedy strategy, can entirely deplete the resource. When a player can entirely deplete a resource by employing a greedy strategy, the game is said to have a low threshold.

> **BET ON IT**
>
> The Tragedy of the Commons model applies to more than overfishing and deforestation. In California, for example, mismanagement of geothermal energy sources have resulted in their depletion, and in Australia, water consumption by farmers is now restricted to prevent worsening of drought conditions that would result if the farmers siphoned off as much as they wanted. Another example involves carbon emissions and global warming: every time you drive your car when you could get there by more eco-friendly means, you are employing the greedy strategy.

Keep in mind that activities like fishing and forestry are repeated over long periods of time. Fish in the North Atlantic and forests in South America do not disappear overnight. So although we are modeling this behavior with a single play of a game, it should be understood that the two strategies M and G refer to long-term behavior.

Another issue is that the players might have different time horizons. A player who harvests modestly would appear to understand that the payoffs occur over the long term, while a player who seizes everything in sight is behaving myopically. Perhaps the greedy players truly don't believe that they are ruining conditions for future generations.

The situation calls for us to describe the players' utilities asymmetrically—according to how *they* see the situation rather than how we see it. A player who is greedy may well have a different utility function than one who acts modestly.

POISONED PAWN

Don't be too quick to assume that greedy play is necessarily immoral. It might be the case that the greedy players simply can't afford the economic consequences of adhering to production limits on activities like fishing and foresting.

With just two players, the Tragedy of the Commons sounds like a Prisoner's Dilemma. If one player plays M, the other player's greed is highly profitable (as measured by the greedy player). If they both play G then they share a large catch in the short term, while if they both play M they have smaller short-term profits but leave some of the resource around to regenerate itself for the future.

The following matrix represents the two-person Tragedy of the Commons as described.

Tragedy of the Commons, two players

	M	G
M	3, 3	1, 4
G	4, 1	2.5, 2.5

Low-threshold Tragedy of the Commons with two players.

This game differs from the "usual" Prisoner's Dilemma, or PD, in some ways. The biggest difference is that if at least one player is greedy, the two players end up sharing the entire resource at once. The mutually modest payoff of 3,3 exceeds the total of 5 points as summed in the other entries because this forward-looking pair of strategies does not wipe out all of the resource. The game is still a PD since it satisfies the defining conditions.

What about with more than two players? First of all, depicting a three-person game in strategic form requires a new trick, which I carry out in the following diagram. But aside from that, the key feature to understand is whether it takes just one greedy player to entirely destroy the resource, or whether it takes two or more greedy players to do so.

For now, let's continue with the low-threshold assumption that a single greedy player can get a large take and, by doing so, clean everything out. Continue to assume that the total amount of the resource is 5 units, except when all players play modestly. The following pair of matrices represents the transition from two to three players.

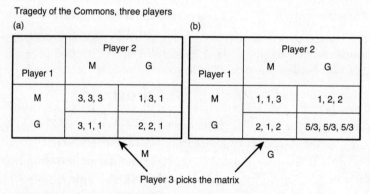

Low-threshold Tragedy of the Commons with three players.

Each matrix entry has three payoffs (to players 1, 2, and 3, respectively). Player 1 picks the row and Player 2 picks the column as before. Player 3 picks the matrix: if Player 3 plays M, they all end up in matrix (a), while if Player 3 plays G, they all end up in matrix (b).

Now let's analyze the game. Suppose Player 3 plays M. Remember that Player 2 receives the middle payoff in each cell. Notice how strategy G in column 2 weakly dominates M in column 1 (if Player 1 plays M, the G payoff of 3 for Player 2 equals their M payoff of 3. But if Player 1 plays G, then Player 2 is strictly better off playing G because the corresponding payoff of 2 is better than 1 unit from playing M).

The same reasoning, although with different payoffs, holds for matrix (b). This means that no matter what players 1 and 3 do, Player 2 is better off playing G.

If we continue with this logic for Player 1, it turns out that in each matrix, strategy G for Player 1 weakly dominates strategy M. Therefore, both players 1 and 2 will play G.

What about Player 3? This player always receives the third payoff in each cell of either matrix. Notice now that in every cell, Player 3 is at least as well off in matrix (b) as in matrix (a). This means that strategy G for Player 3 weakly dominates strategy M.

Finally, let's examine the payoff amounts: do they preserve the idea of the Prisoner's Dilemma? The answer is yes. In matrix (a), the temptation of 3 is at least as good as the mutually cooperative payoff of 3, while the next best payoff is mutual defection and the worst is the sucker's payoff. So matrix (a) is a PD. Matrix (b) is not a PD, since Player 3's greed will reduce the value of mutual cooperation by the others. However, the final outcome is the same as a PD, and the message is still that mutual greed will carry the day.

We've just determined that if all it takes is one greedy player to deplete the common resource, then the multiplayer scenario plays out like a PD, at least given some reasonable assumptions about the resulting payoffs. But what would happen if it takes more than one player to wipe out the common resource?

The Commons Game with a High Threshold

Let's consider one more important variation of the Tragedy of the Commons. Think about what happens when we have three or more players, and a threshold number k. With at least k greedy players, the common resource is entirely depleted, and the greedy players get a bigger share than the modest players. With less than k greedy players, the resource is mostly preserved and, because of this, the sucker's payoff is not so severe.

> **BET ON IT**
>
> You're likely to see some variation in how the Tragedy of the Commons is described. Some game theorists declare that the Tragedy of the Commons is simply a multiplayer PD. Others formulate the game a bit differently and conclude that the game is not a PD. I have been careful to state the assumptions and formulate the game in two different ways. Some Commons games in the real world may behave slightly differently from others. The important thing is to avoid a blanket classification and instead understand how particular subtleties may differ somewhat from situation to situation.

Here's a representative situation. Suppose there are three players and the threshold is 2. This means that if two or three players are greedy, the resource will be wiped out; but with no more than one greedy player, the resource is preserved. Using the same base total of 5 for the resource, the following table provides the payoffs to one random player we single out.

Tragedy of the Commons, high threshold condition

	All others are greedy	One other is greedy	All others are modest
M	1	2	3
G	1.67	2	4

- Resource base level is 5 units
- Player plays M or G
- Threshold = 2

High-threshold Tragedy of the Commons with three players.

When all three players are greedy, they wipe out the common resource and each obtains a payoff of 1.67 (= $5/3$) as before. If the third player plays G and there is exactly one other greedy player, the threshold is crossed and the payoff is 2 out of the 5 units, as before. But if the third player is greedy and all others are modest, the resource is replenished and the payoff is highest, at 4 units. Once again, in these cases strategy G weakly dominates strategy M.

As already mentioned, resources like fishing stock and timber seem to fit the Tragedy of the Commons model. Overfishing has made some species of fish almost disappear from the oceans, and vast deforestation, especially in parts of South America, has been taking place for years. Clearly, what we term greedy behavior has been depleting these resources. What, if anything, can be done about it?

Ways Out of the Dilemma

The Tragedy of the Commons is a classic example of friction between individuals' self-interest and the good of the group. In the low-threshold situation all it took was one selfish player to deplete the resource forever. Such a player may not realize the long-term implications of his actions, or he may be economically unable to rein in his behavior. He might also be aware of the impact of his actions but simply choose to be selfish. No matter what the reason behind his actions, the greedy player rakes in a lot of profit while all others hold back with little to show for their restraint.

In the high-threshold case, it takes more than a single selfish player to destroy the resource. In these cases, the fact that multiple players are necessary to destroy the resource introduces *moral hazard*. A player can still play the greedy strategy and then blame others for any future suffering, since the responsibility to "do the right thing" is now diffused.

DEFINITION

Moral hazard occurs when a particular condition can provide incentive for an individual to behave differently than they normally would. Typically, the deviation in behavior is contrary to societal norms.

Some of the ways to avoid the Tragedy of the Commons are similar to the prescriptions we saw for the Prisoner's Dilemma in Chapter 5.

Recall how in the cigarette advertising game, the mutual defection strategy (analogous to the greedy strategy here) engaged the tobacco firms in a costly television advertising war. But when the U.S. government halted TV ads, the tobacco firms were forced, in essence, to cooperate. Playing this dominated (cooperative) strategy in turn led to higher profits. How can this be arranged in the Commons situation?

Consider a number of farmers who compete not only in the marketplace, but also for the same limited water source to irrigate their crops. Each farmer is tempted to draw off abundant quantities of water, but knows that if others followed suit, drought conditions would result.

The farmers might go on living and working in an atmosphere of great suspicion. But imagine that one day, a bold visionary recruits other farmers to march on the government. Their demand is that the government enacts legislation to limit water intake to individual farms. Suppose they actually have the technology to verify that the newly established limits are enforced. Each farmer's short-term income might take a hit but their businesses will live to see another day—probably many more.

This strategy requires the farmers to recast their competitors as potential allies. Such thinking is at the heart of what game theorists Adam Brandenburger and Barry Nalebuff call "Coopetition."

In the Tragedy of the Commons, the unfortunate thing is that even when most players act with restraint, a few bad apples end up spoiling the resource for everyone. Now let's tweak the concept in the following way: we'll accept that most players will let greed get the better of them. But what if it took just one player's admirable restraint to avoid complete disaster?

> **NO BLUFFING**
>
> Brandenburger and Nalebuff describe how managers can expand their notions about competition to their advantage. The key is to turn certain competitive situations into cooperative ones. The farmers compete in the marketplace—and they should—but they shouldn't compete for water, because they can all lose. Very often changing the rules, changing the game, or changing the way you look at the competition can make all the difference.

The Volunteer's Dilemma

The Volunteer's Dilemma is simply this: who is going to sacrifice herself for the good of the group? In some wartime situations, the sacrifice is indeed giving one's life so others can survive. In more peaceful situations the sacrifice need not be extreme. In a sense, the Volunteer's Dilemma is the mirror image of the Tragedy of the Commons and, while less important economically, it is psychologically more interesting.

There are well-documented situations in which people have given up their lives for others, and these occurrences have spurred a great deal of research and reflection into the nature of altruism. But this is not what we're after just now. Instead, let's investigate the milder cases where the sacrifice involved is, for example, monetary, and not too harsh.

In the mid-1980s a popular science magazine used a contest to gain insight into restrained, altruistic behavior. *Science 84* gave readers the option of sending in postcards asking for $20 or $100. The magazine offered to pay people the amount they asked for provided that the proportion of those asking for $100 was below 20 percent. Otherwise no one would get anything. In this case well over 20 percent of the respondents asked for $100 so the magazine didn't have to dig into its coffers after all.

A more scientific study was published in 1988 by Anatol Rapoport. Rapoport set out to investigate behavior in a number of situations called *social traps*. The details of his Volunteer's Dilemma study were as follows:

- Each subject could select one of two strategies: C for cooperation and N for noncooperation.
- All subjects picking C receive $5 and all subjects picking N receive $10, provided that at least one subject picks C. If no one picks C, then all subjects get a payment of $0.

> **DEFINITION**
>
> A **social trap** is a situation in which individually rational decisions by all members of a group result in an outcome that is not efficient for the group.

The following matrix represents the Volunteer's Dilemma. Each of the two rows represents a strategy for a randomly chosen player. The two columns show the payoffs in dollars to the single player only, depending on the other players' selected strategies.

Volunteer's Dilemma

	At least one cooperator	No other cooperators
C	5	5
N	10	0

There are no dominant strategies in this game. But be careful not to interpret the game as an ordinary nonzero sum game; the columns do not represent strategies for an opponent but are aggregated behaviors over a subset of other players. You can certainly expect things to be different when just one other player is present as opposed to a thousand other players. (In the latter case you would hope for, and expect, at least one "nice guy" to fall on his sword, so to speak. Rapoport, however, pointed out that this logic might lead to a smaller likelihood of cooperative behavior since everyone might think this way and therefore not cooperate.)

I revisit Rapoport's study of the Volunteer's Dilemma in the final section of this chapter. But first, let's consider yet another game in which there is a clash between what's best for the individual and what's best for the group.

The Free Rider Problem

The Free Rider problem refers to situations where people are supposed to contribute toward the use of a certain good or service, but when payment is not enforced, some users don't contribute. An example of the Free Rider problem is when someone hops a train and gets to her destination before the conductor can charge her for her fare. But free riding occurs in many other circumstances as well; for example, when people listen to public radio without ever making a contribution, or when street performers start to pass the hat around and the crowd disperses like nobody's business.

As soon as we realize that individually rational behavior will lead people to free ride, the problem begins to sound ominous. We might expect many important institutions in our society to be endangered, but whether this is true is unclear.

Although the Free Rider problem is similar to the Tragedy of the Commons and the Volunteer's Dilemma, it differs from them in important ways. In the Tragedy of the Commons a single player or a small subset of players can wreak irreparable harm on a community by their selfish behavior; in the Free Rider problem, larger numbers of players can take advantage of the good graces of others without causing an impending, or at least a permanent, disaster.

In the Volunteer's Dilemma, we usually focus on the one poor individual who makes the sacrifice; in the Free Rider problem, there are usually numerous free riders, although they tend to be in the minority.

The following matrix shows a small example with two players. Here, the two players decide simultaneously whether to contribute toward a public good (this can be a library, radio station, etc.). The good will be provided if at least one of them contributes. The cost of contributing is c dollars and each of them measures the benefit at $100. Let's assume that c is less than $100.

	Contribute	Don't Contribute
Contribute	100-c, 100-c	100-c, 100
Don't Contribute	100, 100-c	0,0

The Free Rider problem two players.

There are two pure strategy equilibria in this game, in which one person contributes and the other doesn't. Such an outcome is not very satisfactory. There is also a mixed-strategy Nash equilibrium. We won't solve for it because the true application typically involves a set of players in the hundreds or thousands (though, of course, some Free Rider problems involve just a few roommates and a stack of dirty dishes to be washed). In such a large game it is not clear that the intuition from an equilibrium strategy with two players would carry over to the larger arena.

Although we are not following through with an equilibrium analysis, it is reasonable to assume that with many players, at least some will contribute, given the mixed-strategy nature of the game. Stepping back a level, surely some players will reason that they can safely free ride on the expected contributions of others. This is a discouraging revelation.

Some Experimental Results

Let's find out what happened in Rapoport's experiment on the Volunteer's Dilemma. First, Rapoport wanted to know whether players who had more knowledge about the game played differently from those who did not. Some players were given full instructions, which explained the nature of the social trap and how to obtain a group rational solution, while other players were told the rules of the game and nothing more.

In addition to the full and minimal instruction groups, Rapoport ran three other variations of the dilemma, each with a hypothetical narrative, and sorted the subjects into different categories depending on their assigned role in the story.

Rapoport's five categories were professionals; students; business people; small business employees; and visitors to the Ontario Science Centre. There was some variance in behavior across these subjects, but the more important results were that 41 percent of the subjects with minimal instructions cooperated, while 52 percent of those with full instructions cooperated.

While the results may seem mildly encouraging—almost half of all subjects chose the cooperative strategy—we can't make too much of the issue, since the subjects were agreeable enough to take part in the study to begin with.

Positive experimental results in the Volunteer's Dilemma are important to note. However, you may remain worried about the likelihood for global economic and social instability predicted by the Tragedy of the Commons. It turns out that in this sphere we have more reason to be optimistic than the Commons analyses provide.

Nobel Laureate Elinor Ostrom has shown that actual behavior in Commons situations is not as dire as theory would predict. In fact, in a number of studies in the 1980s and 1990s she found that local users of a common resource often found ways to cope with the problem of perpetuating the resource. The levels of cooperation in the real world, while not ideal, are substantially higher than the game models predict.

Interestingly, in Ostrom's findings, government agencies were frequently unsuccessful in their attempts to manage the resource; rather, the local groups would typically self-organize as they moved toward a resolution. In other words, Hardin's original solutions discussed at the beginning of this chapter—to either regulate or privatize the resource—are not the only, and perhaps are not the best, policies to employ.

Finally, what about the Free Rider problem? Does it live up to its discouraging looking state of affairs?

Clearly, in the real world there are free riders. But does this doom the various honor systems that have been in place for generations?

As with the Volunteer's Dilemma and Tragedy of the Commons, there is reason to believe that the Free Rider problem is not as harmful as it would appear. Notably, the English band Radiohead offered an album as a pay-what-you-want download in 2007, with great success. Though not the first instance of this pricing model, the band's visibility ensured that business gurus would give some serious thought to pay-what-you-want as a viable commercial model.

There are plenty of other instances where an honors system, even in our cynical times, has proven generally effective as a way to charge for goods. Some people will free ride, but the takeaway is that a surprising amount of the time people contribute enough to support the system.

The Least You Need to Know

- The Tragedy of the Commons is usually structured as a multiplayer PD.
- The Volunteer's Dilemma in strategic form has no dominant strategy.
- The Free Rider problem, the Tragedy of the Commons, and the Volunteer's Dilemma are all related.
- The standard solutions to the Tragedy of the Commons are either to regulate or privatize the resource.
- Despite the PD nature of the Commons game, real-life instances show much more cooperation than theory would predict.

Auctions and Eliciting Values

Chapter 15

In This Chapter

- The purpose of auctions
- Different auction features
- Eliciting honest valuations
- Auction theory and practice

Auctions offer interesting insights into many aspects of our economy and culture. For one thing, after the Internet became a pivotal force in our lives, auctions have taken off in the commercial sector as well as in our personal spheres. In the context of game theory, auctions are nothing if not filled with strategic behavior among multiple players. Perhaps most importantly, auctions provide a fundamental way to elicit our valuations of things.

In this chapter I introduce the most common types of auctions and the strategic behaviors and group consequences associated with them. I walk you through how auctions can be designed to elicit honest valuations from bidders. Finally, I describe a couple of important examples from auction practice.

Types of Auctions

Auctions come in many forms, but all of them have one feature in common: there is at least one item of interest, and different players compete through a bidding process either to buy or sell that item. The player with the best bid—the highest bid when competing to buy, or the lowest bid when competing to sell—is the winner. All the other players lose.

> **BET ON IT**
>
> Virtually all auctions feature asymmetric information; players may know something about what the item is worth to themselves, but they don't know what it is worth to others. In the language of incomplete information, players may know their own type, but they do not know any other player's type.

Let's first establish what auctions have to do with eliciting valuations of things. Suppose you have a painting—unfortunately, not a Rembrandt—that you wish to sell. One way to sell it is to have a garage sale. You might put an optimistic price tag of, say $100, on it. Perhaps some passersby get interested in it and maybe someone buys it. For all you know, the person who bought it might have paid $150. You'll never find out. But had you asked $150 in the first place, maybe no one would have bought it at all.

Another strategy is not to have a price tag. When someone gets interested, you ask them what they're willing to pay. But in the absence of competition, few people will reveal their maximum valuation, and you won't take in as much as you could have. The point of auctions is that they are competitive processes that force the bidders to pay more than they might have otherwise.

Although the end result of an auction is to match buyers and sellers, the results may still impact the other members of the larger group. In the usual situations the losers in the auction are monetarily no worse off than before. The economics textbooks treat them as having a net outcome of zero; but this does not account for the sting we feel when we lose a desired item at the silent auction for the local school or get outbid at the last second on eBay.

Let's examine the different types of auctions, and consider the strategic behavior that is characteristic in these varied setups. We'll assume at first that an auction applies to just one item to be sold. I use the terms bidders and players interchangeably.

First-Price Auctions

There are actually two common types of auctions in this category. Perhaps the most typical auction—the kind you would see on TV—is the ascending-bid, or English, auction. In these auctions, either an auctioneer calls out the prices or the bidders raise the amounts themselves. The price keeps increasing and bidders drop out until only one is left. That bidder wins the object at the last announced price. The others walk away empty-handed.

Ascending-bid auctions can get exciting when everyone is in the same room and there are at least a couple of bidders who outdo each other until all but one of them give up. There are, however, other ways to run an ascending-bid auction. For example, prices can appear on a screen while the bidders, perhaps at remote locations, submit bids electronically.

A different way to run a first-price auction is by having the bidders submit sealed bids. This approach is more discreet: each bidder writes down his or her bid privately and seals it in an envelope to be given to the auction manager. The bidders may not even know the identities of the other players. Each player bids once. As before, the highest bid is the winning one.

It may strike you that the ascending-price auction, especially the exciting kind where the bidders are face-to-face and desperate to outdo each other, would generate higher bids than the sedate-sounding sealed-bid variety. But this is not necessarily true.

In the ascending-price auction, the different players have an opportunity to gauge the others' valuations from their behavior. Some bidders may show some signs of recklessness or reluctance and others may drop out, believing that they don't have the resources to match the more aggressive bidders. When players drop out, the winning bidder might win at a price well below his or her maximum.

On the other hand, in a sealed-bid auction, there's only one chance to bid. This generates a lot of uncertainty as to what the others will do. Many bidders are risk-averse, and here's how it affects them psychologically: they worry that a low bid will cost them a desired item, and to avoid losing out in this fashion they make a relatively high bid. They'd rather win and pay almost full price than try to "cheat" and end up empty-handed and regretful. All in all, it's hard to say whether ascending-price beats sealed-bid.

A natural question at this juncture is to ask what strategy a bidder should follow in a first-price auction. Let's assume that each bidder knows how much the item is worth to them. One obvious rule is not to bid more than the item is worth. If you bid too high and actually win the item, you have to pay the full amount, leaving yourself with a net loss.

The next realization is that bidders should, in general, not bid as much as the item is worth to them. If you bid the exact amount the item is worth to you, and end up winning, you might have won the item for a smaller amount. So a general rule would seem to be that bidders should *shade* their bids.

DEFINITION

To **shade** a bid is to bid less than one's full valuation for an item. (Note: some people say *shave* instead of *shade*.)

Of course, these rules are sometimes hard to follow. In ascending-bid auctions, the desire to beat the competition often heats us up emotionally, which confounds our bidding strategies. And in sealed-bid auctions, as I've said, risk-aversion and anticipated regret may shift our strategy, too.

Let's look at an example of the advantages of shading your bid in a sealed-bid auction. To keep things simple, let's suppose the item for sale is worth $20,000 to you. Suppose there is only one competitor, and you are pretty sure that the item is worth between $15,000 and $25,000 to him, with his valuation evenly distributed within that interval. Let's suppose that he bids his actual valuation.

If you bid honestly at $20,000, you will win the item half the time, because $20,000 is exactly halfway along the $15,000 to $25,000 interval for the other player. But now think about offering only $19,000 instead. Because the other bidder's valuation is, probability-wise, spread evenly between $15,000 and $25,000, then half the time he would have offered more than $20,000 and you would have lost anyway. So half the time, it won't matter whether you bid $20,000 or $19,000.

But 40 percent of the time, his bid would have been between $15,000 and $19,000. (From 15 to 19 is 40 percent of the interval from 15 to 25.) In this case, you win with either the honest $20,000 bid or with the lower bid of $19,000. In other words, 40 percent of the time you save $1,000 by shading your bid.

Another 10 percent of the time, however, you get into trouble. Here, your rival has bid something between $19,000 and $20,000. You lose by shading your bid in this case. But the amount you forego is way less than $1,000, since the item is worth $20,000 to you and you would have paid exactly $20,000. What you're giving up in this case is not much of a deal.

This probability distribution is shown in the following table.

Rival Bidder's Valuation

Rival Valuation ($K)	Proportion of time
20 – 25	0.50
19 – 20	0.10
15 – 19	0.40

The other bidder will bid between $15,000 and $25,000 with equal probability throughout this interval.

Putting this all together, by shading your bid from $20,000 to $19,000, 40 percent of the time you gain $1,000 while 10 percent of the time you lose a tiny amount. Therefore, you are better off shading your bid.

BET ON IT

Can you lower your average payout by shading your bid even more? What if you bid $18,000 instead of $20,000? What about going down to $17,000 or just $16,000? It turns out that by doing some more advanced math, or just using trial-and-error, we find that the optimal strategy is to bid $17,500 in this case.

Two things to realize: first, this analysis only included one competing player. As the number of bidders increases, the increased competition will cause you to shade your bid less and less. Secondly, the other bidders will think the same way you do, so it is fair to conclude that everyone in a first-price sealed-bid auction will shade their bids, but less so as the number of players grows.

The foregoing analysis must make the seller worry a little bit. Naturally, the seller will try to discover a way to minimize the potential losses that result from bid-shading. Keep this in the back of your mind while we explore other varieties of auctions.

Second-Price Auctions

In a second-price auction, everyone makes a single, sealed bid. The player with the highest bid wins the item, but will only pay the second-highest amount bid for it. This sounds like a good deal for the bidders and a lousy deal for the seller.

There's one thing to point out, however. Let's consider an ascending-price auction run in the following way: the price keeps rising continuously and, as it does, the bidders drop out. Suppose there are just two bidders left. The price continues to rise. What happens at the moment when the second-highest bidder drops out? The highest bidder will win the item at the exact price where the rival dropped out. The price will not continue to rise all the way up to the winner's valuation.

In this case the ascending-bid auction yields the same result as the second-price auction. So in second-price auctions, the sellers don't necessarily have to worry so much about getting a lower price than with a first-price auction. Nevertheless, the second-price format is less commonly used but, as you will see, it plays a very important role in auction analysis.

Common-Value Auctions

In common-value auctions, the bidders do not even know what the item is worth to themselves, let alone what it is worth to others. Common-value auctions, because of the lack of information, are the trickiest ones to get involved with.

When I describe the item for sale as being of unknown value, I do not mean that it is a particular van Gogh but you are not sure how much it's worth to you. I mean that the item is something like an offshore oil field of very indeterminate value. If the government wanted to raise money as well as increase the oil supply, it might use a first-price, sealed-bid auction to provide the rights to drill in such an offshore site.

Prior to the auction, each bidder will have some private information about the site. This information would be obtained by geologists employed by the oil companies that plan to bid for the drilling rights. But different geologists, exploring different areas of the site, may turn up different data.

No set of data yields anything like a perfect forecast of underground capacity, so the different companies are not only uncertain about what their own data says, but realize that the other companies' data may be entirely different. This level of uncertainty leaves bidders vulnerable to what is called the *winner's curse*.

> **DEFINITION**
>
> The **winner's curse** occurs when the winning bidder discovers that the item they won is not actually worth as much as they paid for it.

Because so much is unknown about the item in a common-value auction, it is difficult to assess its value and even more difficult to strategize about bidding, given you have no idea what the item is worth to anybody. One strategy, though, should be apparent: since the uncertainty and therefore the risks are great, bidders will strongly shade their bids in order to avoid the winner's curse.

Other Auction Types

It's nice to get some exposure to other kinds of auctions even though we won't go into any depth with them. One oft-used auction model is the Dutch auction. This auction uses decreasing prices to sell the item. The initial price is very high, presumed to be way out of reach of all bidders. The price then comes down, either as announced by an auctioneer or on a monitor, and as soon as someone indicates they will buy the item, that bidder wins the item at that price.

To get an idea about the increasing sophistication of auction design in recent years, consider the so-called Anglo-Dutch auction as developed by economist Paul Klemperer. It involves two stages. In the first stage an English auction is used to gather the highest bidders. The top bidders are selected for the next round.

At the second stage, the finalists each submit a sealed bid. The top sealed bid (or bids, if there are multiple items) is the winner. In an example described later in this chapter, the winners paid a price lower than their original bids. You might have expected a greedier-sounding second-stage but this example illustrates why a successful auction might undercharge the winning bidders.

Vickrey's Insight

Although auctions have been around for thousands of years, economists failed to study them using the tools of game theory until Nobel Laureate William Vickrey published a remarkable paper in 1961. Vickrey was able to find a way to deter buyers from shading their bids.

Using a sealed-bid auction, Vickrey's master stroke was to let the winner only have to pay the second-highest price. The idea is that if the winning bidder paid the second-highest price, and not their own higher offer, they wouldn't need to shade their bid. In other words, bidders would have no incentive to manipulate the system. What is remarkable is that this nice idea is not just wishful thinking; Vickrey showed that bidding one's true valuation is a dominant strategy.

BET ON IT

Second-price auctions were developed by Vickrey, and are also called Vickrey auctions. They remain a powerful force in auction theory and play a role in auction practice, especially in Internet applications.

Here's how the argument goes. Suppose you are bidding in a second-price auction and your true valuation of the item is equal to $100 (just to have an arbitrary number to work with). It won't help you to overbid, and you might do worse by doing so. For example, if you bid, say, $105, there is a chance that someone else has bid $104, and in that case you end up paying $104 for an item you value at $100.

What if you shade your bid? If someone else had outbid you to begin with, by shading your bid you would lose anyway. But suppose the second highest bid is, for example, $96. Bidding $99 or $98 will still secure the item for you at the price of $96. But you wouldn't gain anything by shading the bid. However, if you shade your bid below $96, the player who bid $96 will win the item. This means you will lose the item that you would have made a profit on.

In other words, whether you inflate your bid or shade your bid, it will either make no difference or will land you in trouble. Deviating from your true valuation can never be to your advantage; put differently, in a second-price auction, honesty is the best policy!

Now we can see the attractiveness of this auction for the seller. If the buyers are rational, they will all bid honestly, meaning that no one will shade their bid. And we already saw that the outcome of the second-price auction, in terms of the final selling price, is the same as that in an ascending (first-price) auction. This means that the seller need not give up some sort of premium when settling for only the second-highest price.

Vickrey's work opened up a whole field of research in economics and game theory. Researchers realized the power that resided in the way auctions and other economic markets could be designed. If telling the truth is a good thing—and most of us think it is—then why not explore how incentives to do so might be developed for a whole host of economic and social applications?

We have some theory under our belts now, so let's have a look at some results from actual auctions.

Auctions in Practice

The vast literature on auctions contains a mixed bag of results and recommendations. I have included a few choice morsels in the following sections.

Misuse of the Vickrey Auction

Before you get too excited about Vickrey auctions, keep in mind that sometimes things don't work out the way you expect them to. In 1990 the New Zealand government decided to sell the rights to certain radio airwaves. Having digested some auction theory, they selected the second-price auction as the way to generate revenue. But in a couple of cases, there was only one real bidder, and the second-highest price was nowhere near the highest bid. In one instance the winning bid was NZ$100,000 but they ended up paying just NZ$6 (the second-highest bid!) and in another one a NZ$7,000,000 bidder paid a mere NZ$5,000. What a loss, not to mention a huge embarrassment, for the government.

The lesson for sellers is to ensure that there is a guaranteed minimum price. Secondarily, from a seller's perspective, make sure the auction market isn't too thin. The more potential buyers, the better.

Awarding TV Rights

Whenever exclusive coverage to a major event goes to a particular television network, I'm sure we all tacitly understand that the winning network must have offered the best deal. But how are those deals reached, and what do auctions have to do with it?

Game theorist John McMillan presented a study on auctioning the rights to televise the Olympic Games in the United States. One striking example involved bidding to televise the 1980 Summer Olympics in Moscow.

Often in this book we have relied on games having a certain set of rules that the players must follow. If I am moving a knife across a cake to cut it, actions like you shoving my hand just before you announce "stop" are out of bounds for us. In the auction setting, if we agree on a sealed-bid auction, we assume that no one peeks inside another's envelope.

However, in the auction for the U.S. TV rights for the 1980 Summer Games, the host country did some interesting things in order to get the bidding to take off. ABC was awarded the contract for the 1976 Games before the other networks even submitted a bid. The Soviets were determined not to let that happen in 1980, and they tried a number of different tactics that turned out to be very effective, fetching them well over double, in real (inflation-adjusted) dollars, what the Canadians received in the previous Summer Games held in Montreal in 1976.

According to McMillan, the Soviets first made an initial demand that was triple what they expected to get. This gambit may remind you of how certain merchants open their bargaining negotiations. Such a demand could be flatly ignored, but in some negotiations it might actually influence the bidders.

Next, the Soviets met with the network representatives both together and separately. The networks knew they were being played off against each other, but once the fighting started it was impossible for them to withdraw and demand a set of rigid rules to proceed with. Ostensibly, the negotiation was run as a sealed-bid auction, but the Soviets would share some of the private bids with the other networks. They would eventually announce a winner, give the others 24 hours to respond with a better offer, and then repeat it all over again.

> **POISONED PAWN**
>
> When participating in an auction, make sure that the ground rules clearly describe how the auction will be run. Further, make sure there is a mechanism in place to deal with someone breaking the rules. The Soviets knew in the 1980 Games that they would be able to get away with certain tactics because they didn't foresee much future dealing with the American networks, and the networks had little recourse to legal action.

Have I just described an auction, or something out of the Wild West? The lesson seems to be that the players need to have a fixed set of rules to play by. However, if there is no retribution when the rules are broken, don't expect everyone to play by them.

Auction Successes

Since 1994 the Federal Communications Commission (FCC) has conducted a series of so-called spectrum auctions to sell licenses to wireless providers. These auctions raise billions of dollars for the government (the 700 MHz auction in 2008 raised almost $20 billion) and have proved to be a superior way of doing business as opposed to other ways of awarding such contracts.

At first glance auctions might seem to be the best vehicle with which to award licenses and contracts such as spectrum frequencies used by wireless companies. But prior to 1994, contracts like this were awarded by either lottery or through other types of dealing, called *beauty contests* in the literature.

> **DEFINITION**
>
> A **beauty contest,** in economics or game theory, usually refers to one of two things. In the present context, it's a process where decision makers award a contract after viewing a number of presentations or proposals and picking the "best-looking" one. In other contexts, it means a decision maker's choice of the "best" alternative is the one he or she believes others think is best.

Giving out licenses by lottery provides no particular way to maximize revenue for the seller or arrange a good match between the bidders and the items. Beauty contests—in this case, evaluating proposals—are suspect because they leave decision makers subject to favoritism, if not corruption, and they lack transparency. There's no way for anyone to know whether the winner was truly deserving.

Most of the FCC auctions sell a number of different licenses at once. The different licenses apply, for example, to different frequency bands or to different geographical areas. How should the FCC design these auctions? There are many different issues to think about. First, which basic auction design should be employed:

- Ascending first-price auction
- First-price sealed-bid auction
- Second-price auction

There are many other points to decide. Let me go through some of the more important ones. Should the FCC auction off the different licenses separately or should they allow bidders to bid for various combinations of licenses? The San Diego market might be more attractive to a provider if they already locked up a license for Los Angeles, for instance.

What about timing? Should the FCC run the auctions at different times (perhaps in a particular sequence), or simultaneously? And if they run the auctions simultaneously, should they break the bidding up into stages or have the bids generated continuously?

On top of all that, the FCC worries about the possibility of collusion among the different players. Maybe certain low bids convey messages like, "I won't compete for Chicago if you let me get South Florida." Which auction designs discourage or at least limit this behavior?

In the earlier years, the FCC—which is advised by game theorists, as is the fashion in all big government agency auctions around the world—decided not to allow for combination bids. Instead, the different licenses were auctioned separately and simultaneously. The reasoning was that if a bidder appeared to be outgunned for one license, they would be able to make a better bid for a different license. Later on the FCC decided to introduce "package" bidding for groups of licenses.

Deciding between an ascending-price auction and a sealed-bid auction is tough. The sealed-bid structure would probably cut down on potential collusion. And if the bidders are risk-averse, as I have mentioned, they might bid more with sealed bids.

Then again, since the competing companies often have similar cost structures, in the ascending-price auction players would see how much their competitors valued the licenses. This openness might dispel fears of a winner's curse and thus jack up the bids. And even though the FCC charged an entry fee in order to discourage frivolous bidders, maybe the New Zealand experience with the second-price auction was haunting them.

Ultimately, the U.S. government decided to run simultaneous ascending-bid auctions.

Over the past decade 3G wireless auctions have been the rage worldwide. A very successful 3G auction took place in the United Kingdom in 2000. This event, which sold five 3G licenses, raised approximately $34 billion.

The government's objectives, as described by advisors Paul Klemperer and Ken Binmore, were prioritized as follows:

1. to find the best match of wireless providers to licenses
2. to ensure competition
3. to maximize revenue

One way to help reach the second objective was to allow only one license per bidder. But how could the auction design take into account how good the competitors' business plans were, and thus secure the most competent firms?

The argument was that the firms with the best business plans would generally be in a position to value the licenses more, and therefore the government's objective of providing good service would be aligned with the secondary goal of revenue maximization.

NO BLUFFING

Binmore and Klemperer point out that the government could have made much more money by selling just one nationwide license. The winner would become a monopolist and, in the absence of competition, they would be able to generate substantially larger profits than with competitors present. Therefore, the reasoning is that an auction for a single license would result in the different players bidding more than the sum of the bids for the separate licenses.

Another design feature is that the government wanted to encourage new bidders to enter. If new entrants were reluctant to play, then perhaps the industry incumbents would not face competition—and then the auction would fail miserably at revenue procurement.

To encourage fresh bidders, they used the aforementioned Anglo-Dutch auction. The idea was to have an ascending auction for the purpose of generating finalists, so to speak. It would have been hard for incumbents to discourage new entrants to get to this stage. But then the second, "Dutch," stage used a sealed-bid process to further increase bids.

I have run you through this thought process in order to help you understand that there's no one magical auction procedure that works wonders in every situation. It is instructive to see how different objectives or characteristics will lead the auction designer to prefer one setup or another.

Wireless auctions, including those in 2010 for the control of the huge 3G and broadband market in India, have grabbed a lot of attention due to the large price tags. But every day, buyers and suppliers are carrying out frequent, but less heralded, auction deals on the Internet that are improving global supply chain efficiency. Moreover, any time you need something, eBay is just a click away. Auctions have come a long way from a fast talker with a hammer and a block.

The Least You Need to Know

- Auctions are often a better way to award contracts than using a lottery system or reviewing proposals.
- In a first-price auction, shading your bid is usually a good idea.
- Honest bidding is a dominant strategy in a second-price auction.
- No one auction design is always best.

Designing Games for Group Benefit

Chapter **16**

In This Chapter

- Some subtle incentive problems
- An introduction to mechanism design
- Auctionlike mechanisms
- Generalizing the Vickrey approach
- Why no mechanism is perfect

In Chapter 15 you learned how auction design affects the way bidders strategize. Vickrey's insight, in particular, takes strategizing to a higher level by giving us the possibility of designing games in which honest revelation is a dominant strategy. This chapter builds on the theme of shaping participants' behavior in group situations. Game theorists have designed some nifty ways to keep players honest but, as you will discover, these methods all involve some inherent trade-offs.

Overcoming Individual Temptations

Individual greed and its consequences have been a repeated theme in this book. While game theory does not offer a universal solution for all such problems, it has made some headway. We have seen how to avoid, or at least blunt, the horns of the Prisoner's Dilemma, we understand the subtleties of the Tragedy of the Commons and other social traps, and we have learned ways to defuse problems that bridge both cooperative and competitive spheres.

Before exploring games designed to draw out honest revelations from players, it's important to understand a bit more about some subtle dangers that threaten to derail many social and economic situations.

Adverse Selection

In Chapter 7 I worked through the logic of buying lemons—faulty cars—and discovered that when good and bad items are distinguishable only to the seller and pooled together at the same price, potential buyers might get a good deal but are justifiably worried that they'll end up with a lemon. The buyers are scared off because the substantial asymmetry in the types of goods available is kept hidden from them.

Similarly, I pointed out that the very high cost of providing elderly folks with health insurance leads relatively healthy people to opt out while relatively sick people buy in. The result in such an insurance market is that not enough revenue is collected from the healthy individuals to subsidize the huge payout for the less robust. This is why in the United States, for example, Medicare was instituted many years ago to help provide health coverage for the elderly.

Let's think strategically about decision-making in these asymmetric cases. Here are three sticky situations.

- Suppose a used car dealership is stuck with a lemon and wants to sell it. Why should they reveal all the vehicle's flaws and thus sabotage any chance they have of selling it?

- Suppose you are a heavy smoker but still hope to obtain health insurance that doesn't cost, well, an arm and a leg. Will you report your smoking to the insurance company?

- What if you wish to obtain a mortgage but realize that if the banks find out you declared bankruptcy 10 years ago, the loan is a nonstarter? Will you conveniently forget to mention that small detail?

Who can blame smokers for trying to fit in with nonsmokers when it comes time to pay for an insurance policy, or borrowers who believe it is justifiable to put their lending history behind them?

All three examples illustrate *adverse selection*. You have already seen one way to combat this problem: in the lemons market, since the buyer lacks information about the cars, the dealership offers a warranty to ease the buyer's concerns.

DEFINITION

Adverse selection refers to situations where information asymmetry leads to undesirable outcomes. The typical applications are in insurance and banking, where one party is informed of certain risks while the other is not.

In the smoking and bankruptcy cases the individual is informed about his or her condition but the insurance company or bank is not. These firms need a way to combat adverse selection. Their chief weapon is underwriting: trying to discover whether the risk they face is acceptable. Underwriting allows the insurance company (or bank) to discover the private information that you might have preferred to stay hidden.

To summarize, there are two typical ways to lift the fog of asymmetric information. When a seller lacks information about its customers, underwriting can tease out the different types that are under the single umbrella. When the seller has more information than the buyer, a warranty can protect the buyer from the bitter purchase of a lemon.

One other tool can be deployed when grappling with adverse selection: mechanisms that are designed to reveal players' true types—ultimately to the benefit of the community. But first it's important to understand one more subtle influence on individual behavior.

Moral Hazard

In Chapter 14 I introduced the notion of moral hazard in analyzing the Tragedy of the Commons. If it takes multiple players' greed to deplete the common resource, then any one individual can lay the blame at the feet of others.

Moral hazard, though, insinuates its way into situations far beyond the usual Commons scenarios. Suppose you have a really outdated kitchen that you can't afford to update. If somehow, though, it were to become flooded, maybe the ensuing damage will result in the insurance company footing the repair bill and you'll get a decent kitchen out of it. Would you simply pray for that flood or maybe, even unconsciously, do something to bring it about?

What about people who discover that a certain treat is fat-free? They might end up overindulging and taking in more calories than they would have otherwise. And how about those people who used to drive sedately in their compact cars but became maniacs behind the wheel of their SUVs? In these instances a certain condition, like believing that an SUV is safer than a compact vehicle, leads us to adopt risky behavior.

Insurance companies worry about moral hazard because being insured leads many of us to engage—perhaps unintentionally—in behavior that we would not pursue without the insurance coverage. There's one well-known weapon to combat such behavior that goes back at least to Benjamin Franklin: to reduce the chance of an adverse event occurring. Some insurance companies, for example, offer policy owners a rebate if they go to the gym a certain number of times a year.

> **NO BLUFFING**
>
> Ben Franklin helped set up the first American fire insurance company in Philadelphia in 1752. Right after a homeowner purchased an insurance policy, Franklin arranged to have nearby trees trimmed to reduce the chances that the home would catch fire. In the first year of business, Franklin's Philadelphia Contributionship company sold 143 policies, and not one of those properties burned down.

Now that you understand how the temptations of adverse selection and moral hazard can lead people astray, let's turn to a new tool in our fight to keep players honest.

Design for Truthful Revelation

In Chapter 15 I demonstrated how Vickrey's second-price auction gives all potential buyers an incentive to bid their true valuations. The second-price auction works because the actual amount you pay is independent of the amount you bid. Therefore you are not punished in any way for revealing how much the item is really worth to you.

Let's not forget that this part of the book is concerned not just with individual valuations but with group consequences. So now let's consider some examples of interactions in which game theorists have managed to coax players into being straight with one another—not just for honesty's sake but for everybody's benefit.

The schemes that I present provide an introduction to a branch of game theory called mechanism design. Mechanism design is a way to combat problems caused by private information; the idea is to engineer the rules of the game to get the players to behave in a certain way—in this case, providing honest valuations.

Cost-Sharing Schemes

Suppose you are the head of a division in a loosely federated organization that is contemplating an important new technology investment.

Chapter 16: Designing Games for Group Benefit 223

Let's say your organization has five divisions of somewhat different sizes, and that the price tag for this technology is $50 million. You and your four divisional counterparts all believe that the new technology will ultimately be a cost saver but some of the division heads are more excited about it than others.

How do you decide whether it is worth it as an organization to purchase the technology?

You could hire some very high-priced consultants to figure it out for you. But when the smoke clears and they give you their report (which costs almost as much as the technology), you might find that they don't really know your division's ins-and-outs as much as you do.

Next, you might have all of the divisional heads do their own such analysis and then get together to share with each other how much money they will save over the first few years.

Let's think about this approach. What if you believe that the new technology will save your division $25 million. And what if your division turns out to be the one that benefits the most? When it comes down to chipping in for the technology, you might be asked to contribute the most.

Now the game theory lessons start to kick in. If each division is expected to pay some sort of proportional contribution depending on what they stand to gain, then the division chiefs will realize it is not in their best interest to report their savings honestly. Everyone is going to think this way, and if each one underreports their savings, then the group outcome might be very unfortunate: the reported savings may not warrant the technology purchase! Let's see in more detail how this could happen.

Suppose the true savings for the five divisions are as follows:

True Divisional Savings

Division	1	2	3	4	5
Savings ($M)	15	25	7.5	20	17.5

Actual savings per division with the new technology

The total amount saved in just the first few years would be $85 million, while the technology costs only $50 million. Buying the technology looks like a no-brainer. But the wily division chiefs report the following benefits instead.

Reported Divisional Savings

Division	1	2	3	4	5
Reported Savings ($M)	6.5	15	5	9.5	6

Reported savings per division with the new technology

The total reported savings are $42 million. Everyone at the table not only knows that they underreported their own figure, but that everyone else did, too. Nevertheless, no one is willing to budge, and after some glances around the table, the technology proposal is scrapped.

Not all is lost, though. Game theory to the rescue! Suppose that before the meeting you pulled out your copy of *The Complete Idiot's Guide to Game Theory* and flipped to Chapter 16. What you'll find here is a way to elicit honest valuations by using an auction.

Instead of a straightforward auction you decide to run what's called an auctionlike mechanism. Here's how it works. First, the company appoints someone to be the administrator for this procedure. The administrator sets a starting price as might be done in an ascending-price auction. Each player (division head) is asked if they'll pay that price—take it or leave it, yes or no. If any player declines, then that player is excluded from using the technology if it is purchased. (Notice how this prevents free riding.) The administrator sets successively higher prices until either enough money is raised, or everyone has said no.

POISONED PAWN

When you're all on the same team you might forget that the other players still have their own agendas. Chipping in to pay for a profitable new venture sounds very cooperative but underneath the surface each player would still free ride if they could. Auctionlike mechanisms can be very effective because they eliminate the free-rider possibilities.

Here's how you go about setting the price: the cost in this case is $50 million and there are five divisions. This brings us to Step 1.

1. The starting price is set to 50/5 = $10 million. Players privately accept or decline.

The key to the procedure is that any division that does not accept the price at any stage is excluded from using the technology. Each division looks at the current price tag, which right now is $10 million. If the technology will not save the division $10 million, then the rational (and honest) reply is to decline. If they accept, they'd have to cough up the money and that would lead to a net loss.

But if the technology is worth at least $10 million, then the rational response is to accept, because if the player declines, that division will miss out on a profitable opportunity.

Now take another look at the true valuations. Only division 3 finds the $10 million price tag too steep. That player will opt out at Step 1. But despite the fact that all other players accepted, the four of them that remain only contribute a total of $40 million at this point, which is not enough.

Now the administrator raises the price again. This is Step 2:

2. The price increases to 50/4 = $12.5 million. Players privately accept or decline.

Looking at the players' actual savings, we see that each of the four remaining players (#1, #2, #4, and #5) has a net savings that exceeds $12.5 million. If any of them rejects the price of $12.5 million, that division never gets the technology. Since it is rational to be honest and accept, all four divisions do so. Not only is the final result individually rational for all players, but the organization as a whole will benefit from adopting the technology.

This auctionlike mechanism results in divisions 1, 2, 4, and 5 each paying $12.5 million. All receive the technology and will benefit from it. Division 3 will not. Incidentally, suppose one or more of the players had said no in Step 2. Those divisions would have dropped out, leaving some number, k, remaining. The next step would have set the price to 50/k. Continuing in this fashion, the auctionlike mechanism always reaches some conclusion because at some point either all k divisions say yes, or there are none left.

POISONED PAWN

We have hit on a neat way to elicit true valuations so as to benefit the group, but our auctionlike mechanism is not perfect. For one thing, the mechanism is inefficient: it is possible to create examples where the true valuations add up to more than the technology is worth, and yet the mechanism does not fund the technology.

Unfortunately, this process is not perfect. Notice that this mechanism always doles out equal cost shares to all participants. However, as you can see in the example just discussed, the players may end up with very unequal savings. Division 2 has a net savings of $12.5 million (25 – 12.5), while Division 1 only has a net savings of $2.5 million (15 – 12.5).

Finally, the exclusionary feature can be troublesome. While excluding players from free riding is the key to the whole procedure, when the situation calls for everyone to be a user, the mechanism does not work out ideally. In this case Division 3 misses out and there is no foolproof way to bring that player into the fold.

Despite its problems, the auctionlike procedure brings a potent weapon into the game theorist's arsenal.

Vickrey-Clarke-Groves Mechanisms

It's Saturday night and you and your companion are late getting to the movie theater. You absolutely hate to walk into the screening room even a second after the film starts. But there you are, with 10 ticket-buyers ahead of you in the line. Other friends who are already inside have just texted you that the previews are ending.

If you could buy your tickets immediately you'd get inside just in the nick of time. But you can't just cut in front of the line. Or can you? You have an inspiration: you rush up to the person standing at the window and say, "I'll give you 10 dollars if you let us in front of you."

Of course, in real life, we don't do things like this. For one thing, the other nine people you just bypassed might get angry and demand a payment for your cutting in front of them. The person you talked to might simply ignore you. Or they might respond, "Give me a twenty and it's a deal," and then you have to think about whether it's worth $20.

Of course, I am not proposing that we try to design a mechanism so people can swap places waiting in line to see a film. But perhaps there are more important commercial situations where such a mechanism would be useful: when planes are in holding patterns during a flight delay (some might need to land more promptly than others), or when factory orders are backlogged for different customers, some of whom need immediate delivery more desperately than others.

Building on Vickrey's second-price auction, is there some way to design a mechanism for situations like rearranging customers waiting in line, or other more general circumstances in which multiple items are to be bought, sold, or exchanged?

Vickrey's auction result opened up a whole body of research in economics and game theory. In the early 1970s two economists, Edward Clarke and Theodore Groves, independently found a way to extend Vickrey's idea to situations with multiple goods and many players. This may not sound like a big deal, but their methodology in fact applies to a wide class of problems.

To see what Clarke and Groves achieved, consider an auction with several different items and many bidders. The highest bidder for Item 1 wins that item, Item 2 goes to its highest bidder, and so on. Some players might even bid for combinations of items: they can enter a bid for items 1 and 2 together. Recall that the key to Vickrey's procedure is that the amount a winning bidder ends up paying is independent of their bid.

What Clarke and Groves did is to extend the second-price logic for a single item to the more general setting. Let's consider a randomly chosen player, Player P. Player P may make one or more bids. For example, P bids x dollars for one item and y dollars for another item. The following steps comprise the Vickrey-Clarke-Groves mechanism (VCG):

1. VCG receives all bids, including Player P's.
2. VCG calculates the highest total group valuation: this is found from the assignment of items to bids that maximizes the values as stated by the bids. (Player P may or may not have won any of the items.)
3. VCG removes P's bids (as if P never showed up) and recalculates the auction results as in step 2. If P had won any of the items, then deleting P will change the results.
4. VCG charges P for the loss in the value of the assignment that P's bidding causes the rest of the players. In other words, the amount that P will have to pay in the auction is equal to the collective hurt P causes the other players.

The remarkable thing about the VCG procedure is that it is strategyproof, meaning it is a mechanism for which revealing one's true valuation is a dominant strategy for every player (just as truth-telling was a dominant strategy for bidders in the second-price auction).

Furthermore, it turns out that the VCG procedure is efficient in the now familiar economic sense: the assignment of items to bidders that the procedure specifies will always generate the highest total group valuation possible.

> **NO BLUFFING**
>
> Economists Jerry Green and Jean-Jacques Laffont showed that the VCG mechanisms are the only ones that are both strategyproof and efficient. In other words, there is no hope of ever discovering another procedure that provides an incentive for players to always report honest valuations and also yields a socially optimal result.

Let's apply this mechanism to an example. Consider an auction with two equivalent prizes available—dinner for two at a nice restaurant—and four players bidding. We assume for now that the bids represent the true valuations.

Player 1 bids $100, Player 2 bids $90, Player 3 bids $75, and Player 4 bids $120 but wants both prizes and won't settle for just one of them.

Do the Math

It's easy to see that players 1 and 2 are the winners of the two prizes. They are awarded one each. (Together they easily outbid Player 4.) If Player 1 did not participate, then players 2 and 3 would be the prizewinners. The following table lists the valuations to the players other than Player 1 in the auctions when Player 1 either participates or not.

VCG Valuations, Player 1

	Assigned values		
	#2	#3	#4
With #1	90	0	0
Without #1	90	75	0

Values obtained in the auction with and without player 1

Player 4 loses out whether Player 1 participates or not. Since there are two prizes, Player 2 (the second-highest bidder) wins a prize anyway. But it is Player 3 who suffers with the addition of Player 1: Player 3 wins a prize when Player 1 is absent but loses out when Player 1 bids.

Player 1's entry into the game, therefore, causes a net loss of $75 to the other players. This is what Player 1 will be charged.

What about Player 2? We construct a similar valuation table for her:

VCG Valuations, Player 2

	Assigned values		
	#1	#3	#4
With #2	100	0	0
Without #2	100	75	0

Values obtained in the auction with and without player 2

From the table, we can see that Player 2's participation causes equal pain to Player 3—a loss of $75.

Neither Player 3 nor Player 4 needs to be examined because their lower bids ensure that neither has any effect on whether the other players win.

The Analysis

What is the relationship between this procedure and Vickrey's original second-price auction? If, in the preceding example, only one prize was offered instead of two, you can see how the Vickrey auction becomes a special case of VCG. Without Player 1, Player 2 would be the winner, getting a dinner she values at $90. But when Player 1 enters a higher bid, Player 2 loses out on a $90 dinner. Players 3 and 4 are not affected. So Player 1 wins, gets a dinner worth $100, and is charged the $90 of pain (equal to the second price) for knocking Player 2 out of the running.

Wonder why honest bidding is a dominant strategy? A mathematical proof is beyond our mission here. But the pivotal idea is the recurring theme: what any one player pays is independent of their bid. Therefore, manipulating your bid isn't going to affect what you pay, unless you lower it so much that you lose out on a desired item, or you raise it so much that you end up paying more than you'd like.

Internet Advertising

You must wonder sometimes just how Internet search engine companies like Google and Yahoo! make so much money. You probably know that these search engines charge a fee to advertisers to display their links when certain terms are typed into the search box. But how is the fee determined? Not surprisingly, they use a well-designed auction to maximize their revenue.

The actual details of how search engines charge their customers (their "cost per click," and so on) are somewhat complicated. But the following description provides an approximation.

Suppose a number of bidders want their link to appear when an Internet user types "San Francisco" into a search bar. To simplify our example, let's say there are four bidders competing for two sponsored links. Google, for example, uses a sophisticated auction device called a *generalized second-price auction*.

DEFINITION

Generalized second-price auctions have at least one item, and each item's winner pays the amount of the next-highest bid for that item. But in these auctions, because of the way the revenue is generated, honest bidding is not necessarily a dominant strategy.

The four bidders are as follows:

- Player 1: the Hilton hotel chain, bidding for their link, "Hilton Hotels San Francisco"
- Player 2: a dentist who wants you to see his link, "Teeth Whitening San Francisco"
- Player 3: the Hilton family of hotels again, bidding for their link, "Embassy Suites San Francisco"
- Player 4: a tour operator, hoping you get to see his link, "Tour San Francisco by Cable Car"

Of course, just because someone types "San Francisco" and sees a sponsored link doesn't mean she'll click on it; and if she does click, she may not buy the product. But given the millions of searches going on, and thousands of clicks, the advertisers are going to have some success. On average they are willing to pay the search company a small amount for each click.

Suppose the following bids reflect the amounts per click that our four bidders are willing to pay:

- Player 1: $0.45 per click
- Player 2: $0.55 per click

- Player 3: $0.41 per click
- Player 4: $0.34 per click

Based solely on the price-per-click figures, Player 2, followed by Player 1, are the two winners. As it turns out, since people tend to click more on links that are higher up on the screen, the topmost link will go to the highest bidder—"Teeth Whitening San Francisco"—and underneath that will be "Hilton Hotels San Francisco."

> **BET ON IT**
>
> Google and other search engines take into account factors other than the pay-per-click bids. One important determinant of how bids are ranked is a "quality score" that is based in part on how likely it is that a page viewer will click through. (Think expected values: if Advertiser A pays a little more than Advertiser B but B's link is wildly popular while no one ever clicks on A's link, then B is the better choice.)

To ensure that what bidders pay is independent of their own bids, in the generalized second-price auction each winning bidder pays an amount related to the next-highest bid. Honest bidding does not necessarily take place. However, although individual bidders might deviate from truthful revelation, this model appears to provide Google a higher revenue stream than honest bidding would.

In some sense, everyone wins. Advertisers get effective ad placement, Google makes a fortune, and Internet users can find products and services efficiently.

Limitations of Mechanism Design

You've now seen a couple of examples of mechanism design and its promise. But before moving on to the next topic, I have to make sure I don't promise too much. You might recall that the technology-sharing example analyzed earlier in this chapter had some drawbacks. It turns out that these drawbacks are inescapable in mechanism design.

Efficiency vs. Budgeting

You might remember that in the VCG auction example, players needed to make payments to one another to compensate for the losses that each player might cause the others when entering into the game. In auctions, the money to pay for the items, as well as the "transfer" payments, comes from the players themselves. But in some VCG situations the money ends up having to come out of someone else's pocket! Such situations are not *budget balanced*, and their occurrence creates problems for the VCG procedures in general.

DEFINITION

A set of payments made among a group of players together with an administrator is called **budget balanced** if the sum of the payments made by the players is at least as much as what is paid by the administrator. Budget balance means that there is no net deficit.

Here's a quick example of a VCG procedure that is not budget balanced. Suppose there are three players (representing three communities, for example) that are hoping to contribute toward a shared facility. Each player values the facility at $1.5 million. The facility costs $2 million.

Under a VCG scheme, randomly selected Player P pays the total valuation to the other players when P is present minus the total valuation to the other players when P is absent.

With P present, the valuation to the two other players is $3 million; but with P absent, the total valuation is still $3 million. Therefore Player P, and each of the others, will pay zero. But this means that the $2 million to fund the facility must come from somewhere else—the "administrator." In other words, the set of VCG payments is not budget balanced.

This lack of budget balance bursts our bubble, so to speak, and game theorists have tried to work out ways to obtain budget balance in certain classes of situations, or to obtain modified procedures that are almost budget balanced. In other words, budget balance, being important, becomes one of the three main properties mechanisms ought to satisfy; they also ought to be efficient and strategyproof.

It turns out that Green and Laffont showed that no mechanism can be designed that satisfies all three of these conditions. If you need budget balance, you have to give up efficiency or truth-telling. If you want truth-telling you have to give up one of the other two, and so on.

While strategyproof mechanisms are certainly desirable, game theorists can sometimes obtain good results by accepting a slightly weaker version of it. Remember, strategyproof means that every player's dominant strategy is honest revelation. The following condition is almost as good.

What if, in a particular game, the best response to every other player's being truthful is to be truthful yourself? In other words, what if the Nash equilibrium—technically the Bayesian-Nash equilibrium, since players have types—occurs when everyone's strategy is to tell the truth? This condition is called incentive compatibility (IC). The IC property is extremely useful to game theorists because it enables them to obtain other desirable outcomes that strategyproofness rules out.

BET ON IT

Having the concept of incentive compatibility under our belts allows us to peek at the main theoretical result in mechanism design: the revelation principle. The revelation principle (proved by several researchers in the late 1970s, including Nobel Laureate Roger Myerson) states that given a mechanism and any Bayesian-Nash equilibrium of that mechanism, there is a truth-telling equilibrium that yields the same exact outcome. This is huge, because it means that mechanism designers need only search for mechanisms that are truth-telling.

While we see that Vickrey-Clarke-Groves mechanisms provide an ingenious and valuable tool for mechanism design, we also realize that we can't quite get everything we want. Of course, we shouldn't have expected the world. We will conclude this chapter with two more cautionary observations.

Computation and Collusion

To introduce VCG mechanisms I had you think about cutting in front of people in a line to buy movie tickets. To avoid getting into a fight, you were prepared to pay folks to get ahead of them.

Would it be possible to design a VCG mechanism to honestly elicit, from each ticket-buyer in the line, either how much they would pay to move ahead or how much they would have to be paid to move back in the line? The answer is yes. However, to do the analysis thoroughly we would have to take into account all possible permutations, or orderings, of the people in the line.

With some problems, the total number of possibilities can get so astronomically large that even a lightning-fast computer is not going to get through all of them in a reasonable amount of time. And incidentally, there's a theorem in the literature that says if you cut corners and don't count every possibility, then you lose strategyproofness.

In many situations where there are a lot of players, the inherent complexity of the problems can make calculating the VCG payments virtually impossible.

There are many other criticisms that we could level at VCG and other mechanisms, including frequently used auction designs, but the last one I will leave you with involves collusion. In auctions, for example, it is possible that a group of players can arrange that the winning bids are lowball ones and that the prizes are literally a steal.

To see this, consider an auction with two bidders and two items. Player 1 wants item A much more than Player 2, but Player 2 wants item B much more than Player 1. They might agree prior to the auction not to contest each other. The lack of competition will result in the seller being cheated out of much higher prices.

The Least You Need to Know

- Mechanism design is a way to engineer games in order to get around problems of asymmetric information such as adverse selection and moral hazard.
- Auctionlike mechanisms can sometimes solve otherwise intractable cost-sharing problems.
- Vickrey-Clarke-Groves mechanisms induce honest bidding and also provide solutions that maximize group revenue or minimize group cost.
- No rules or procedures exist that are simultaneously strategyproof, efficient, and budget balanced.

Behavior in Games

Part 5

Now it's time to grapple with a thorny issue: game theory prescribes rational action, but it deals with real people, whose brains are wired a certain way. Not surprisingly, people's actual decisions often differ from those given by the theory.

How do we deal with this disconnect between theory and practice? First I review utility theory, which is the classic way to measure how people weigh up alternatives. As you'll see, utility theory is not sufficient to account for all of our idiosyncratic behavior, and something else called prospect theory has come along to better explain our mysterious ways.

In addition to learning how our minds work—with a bit of neuroscience thrown in— this part surveys how people actually behave when playing games in the laboratory. Between prospect theory and the behavioral findings, we find that most people play by the book in some instances. Even more surprising, though, is that when people stray from purely rational strategies, their variations are fairly predictable.

Finally, I demonstrate that when games are played repeatedly, the strategies that people employ are, in a very real sense, *more* rational than the theory prescribes. This super-rationality manifests itself by people's surprising tendency to cooperate with others. Maybe I'm an optimist at heart, but this discovery gives me a good feeling about our future on the planet.

Biology and Games

Chapter 17

In This Chapter

- Game theory in the wild
- Strategies that resist invasion
- How emotions keep us rational
- Games and neuroscience

It isn't immediately evident that other living things play the kinds of games that we study in game theory, and even if they did, you might wonder what there is to be learned that's useful. In this chapter, you not only learn that other creatures are very much involved in the sorts of games people usually think belong just to the human realm, but you also find out, much as in the animal kingdom, that our own biology shapes our strategic interactions.

While game theory purports to study rational decision-making, keep in mind that it's not always possible to eliminate other factors, such as biological impulses, from our strategic palette. Therefore it is important to get an understanding of the origin of some of our tendencies and predilections.

Hawks, Doves, and Their Strategies

Seeing the word "strategies" in the same sentence as a couple of bird species might be enough to launch an objection before I even begin. Why am I using the term *strategies* instead of *behaviors*? I'd mentioned in Chapter 7, and want to re-emphasize here, that strategic action need not be conscious.

To use a bit of poetic license, often we humans act according to our hearts and not our minds. But the point really is that certain behaviors in the animal kingdom seem to work out in calculated ways even though the animals clearly are not doing any conscious computations.

In Chapter 5 I analyzed the game of Chicken, and now in a different context I explore whether the sort of interaction that Chicken embodies is uniquely human or whether it is observed elsewhere in the animal kingdom.

In the 1970s a new breed of biologists was hard at work coming up with mathematical models that would contribute to our understanding of evolution. One of these scientists was John Maynard Smith, who explicitly used the tools of game theory in evolutionary biology and published the first book on the subject in 1982.

Maynard Smith asks us to consider two animals—not necessarily doves or hawks—that are competing for a certain resource, like territory. This resource has value V for either competitor. In game theory values like V usually represent either monetary payouts or else slightly nebulous "utilities," but Maynard Smith explains that what he means is that the so-called Darwinian *fitness* of either animal would be increased by amount V. For example, mastery of the disputed resource by the winner would increase its number of offspring by a certain average amount.

DEFINITION

Fitness in evolutionary biology has to do with the relative ability of an individual to survive and pass on its genes.

To keep things simple, the first approximation to the animals' behavior is to assume that each will adopt one of two strategies. The combatants will either adopt the "Hawk" strategy—to be aggressive and fight for the resource until victorious or injured—or will adopt a "Dove" strategy where it will "display," or pretend to be aggressive to begin with, and then withdraw at the first sign of continued aggressiveness by the opponent.

If both animals pursue the Hawk strategy (and in our sense of game theory, this behavior is indeed a strategy whether or not the animals can understand Shakespeare), one or the other of them will get injured and then abandon the fight. Injury carries with it a cost of C. This cost represents losing a certain amount of fitness.

With a couple of additional assumptions that I will get to next, the game matrix for the Hawk-Dove game can be represented as follows:

The Hawk-Dove game

	Dove	Hawk
Dove	$\frac{1}{2}$V, $\frac{1}{2}$V	0, V
Hawk	V, 0	$\frac{1}{2}$(V-C), $\frac{1}{2}$(V-C)

Payoffs in evolutionary fitness for each Hawk-Dove strategy pair

The assumptions captured in the model are first, if one animal plays Dove while the other plays Hawk, the Dove will retreat at zero cost while the Hawk gains the full value of the resource. If two Doves meet, they share the resource equally. If two Hawks meet, they are assumed to have an equal chance at either winning the resource or being injured.

NO BLUFFING

Biologists work out and test some very sophisticated mathematical models involving foraging, breeding, and territorial behavior. I've started out with a standard model that can be embellished to describe asymmetric contests or situations in which more than two strategies are used in the wild.

Notice that the Hawk-Dove game is nonzero sum. Let's substitute some numbers in for V and C. For example, suppose $V = 2$ and $C = 4$. In this instance fighting is extremely risky, because the cost of being injured far outweighs the fitness gain from winning. We are left with the following payoff matrix.

Hawk-Dove game, Numerical Payoffs

	Dove	Hawk
Dove	1, 1	0, 2
Hawk	2, 0	-1, -1

Payoffs in the Hawk-Dove game when V = 2 and C = 4

The payoff matrix is identical to that of Chicken from Chapter 5. If you'd like, you can try to substitute different values for the two variables V and C and see what the resulting matrix looks like. As long as C is larger than V, there exist the two pure strategy equilibria that you saw in Chicken: (Dove, Hawk) and (Hawk, Dove). In addition to the pure strategy equilibria, the antagonists can also arrive at a mixed strategy equilibrium, although in that case you may wonder how they know to mix their strategies.

Evolutionarily Stable Strategies

Maynard Smith was not satisfied with the ordinary game-theoretic sense of equilibrium. After all, he was dealing with a different world: a domain in which populations are in flux, payoffs measure changes in the gene pool, and different species invade one another's territory. The key question is, depending on the different mixes of strategies used, under what conditions would a population remain stable, and when would a population change?

We know that one way the gene pool can change is through mutation. Suppose there is a population of animals that all use a certain strategy, J. (Strategy J, in our situation, is a mixed strategy.) Now suppose a mutant appears that employs a different strategy, K. When can members of the original population fight off this newcomer? In other words, under what conditions will the existing population, using strategy J, be able to maintain a fitness superiority over strategy K?

Let's let $E[J,J]$ represent the expected value to an animal employing strategy J when it encounters an opponent that uses the same strategy. When the mutant, employing strategy K, invades this population and fights an opponent using strategy J, the expected value for the invader is $E[K,J]$.

Suppose $E[K,J] > E[J,J]$. This would mean that the invader will garner a higher increase in fitness than members of the incumbent population. If this were to happen, then the mutant would be able to successfully invade, destroying the stability of the original population.

This line of reasoning led Maynard Smith and his colleague George R. Price to the following pair of conditions:

For a population to be in equilibrium,

1. Either $E[J,J] > E[K,J]$, or
2. $E[J,J] = E[K,J]$ and $E[J,K] > E[K,K]$.

Condition 1 says that the incumbents rack up more fitness among themselves than the invader does against them. Condition 2 says that if there is a tie in the fitness that J or K obtains against the incumbent animals, then we need to make sure that the incumbents score more fitness when they encounter an invader than the invaders gain against their own kind.

Any strategy satisfying conditions 1 and 2 is called an *evolutionarily stable strategy*, or *ESS*. If these conditions are violated, it means that the invader is able to get a toehold in this particular community and thus will be able to pass its genes on more successfully, at least for the time being, than the incumbents.

> **DEFINITION**
>
> An **evolutionarily stable strategy (ESS)** is a strategy which, if pursued by a population of individuals, will protect them from invasion by others using any alternative strategy.

Let's now look again at the Hawk-Dove game and see whether there is any strategy that fits the ESS conditions. Is a population consisting of all Doves stable? The answer is no: $E[H,D] > E[D,D]$. (In our special case, $E[H,D] = 2$ and $E[D,D] = 1$.) This means that a Hawk can successfully invade a population of all Doves.

But what about Hawks? Can they keep Doves out? In our example, when $V = 2$ and $C = 4$, we have $E[H,H] = -1$ while $E[D,H] = 0$. This means that Doves can successfully invade a population of Hawks. The idea is that the Hawks cause a great deal of injury among themselves while the Doves simply avoid the fight and its costs. As long as $V < C$, we find that neither a population using exclusively a Hawk strategy nor a population using only a Dove strategy is evolutionarily stable.

So the question is, if neither Hawk nor Dove behavior is evolutionarily stable, then what is? The logic of this situation dictates one of two possibilities, if there is an ESS at all: either we must have a mix of animals, some of which are Hawks and some Doves, or there must be some individuals that play mixed strategies: they're sometimes "hawkish" and sometimes "doveish."

> **BET ON IT**
>
> Although having half Hawks and half Doves is very different from having a homogeneous population that plays each strategy half the time, the results, in terms of average fitness gained or lost by the individuals, are exactly the same.

Maynard Smith worked out a number of details that I will skip over, but in conclusion he finds that there are two possibilities for an ESS in this game: either the animals all employ a mixed strategy in which they play Hawk with probability $p = [V/C]$ and play Dove with probability $(1 - p)$; or, there is a mix among the population in which a proportion $p = [V/C]$ play Hawk exclusively and the remainder play Dove. With our values $V = 2$ and $C = 4$, the probability $p = \frac{1}{2}$.

Clearly, we have only just begun to scratch the surface of a rich theory. Maynard Smith and many others worked out many more sophisticated models. What's more, the models hold up in the field; in other words, biologists have been able to test the models by observing wildlife, and the results very often support the theory.

Commitment and Other Virtues

Back in Chapter 8 I presented an intriguing paradox, due to Robert H. Frank, that stems from emotions influencing our decision-making. While emotions may cause us to engage in certain behavior that is not in our short-term best interest, we may end up better off in the long run because of them. The example I gave was the hotheaded desire to seek revenge.

The idea was that your retaliation against an enemy was not in your best interest, when pros and cons are weighed rationally. But your animal instinct—an uncontrollable urge—while costly to you, might prevent some future adversary from crossing you next time.

In this book I've mostly put emotions aside, until now, and focused on purely rational decision-making. One of the key concepts in this regard is subgame perfection—recall the idea that one should disregard equilibria that are based on threats that cannot credibly be carried out.

Now suppose that one player makes a threat—for example, one of retaliation. The other player might see it as irrational, and therefore a bluff. What Frank calls the "commitment problem" is this: how do we convince others that we are indeed committed to a certain behavior—for example, revenge—when it does not appear to be in our self-interest?

What Frank really wants to get at is whether it is possible that our emotions, and sometimes even their physical manifestations, provide an effective aid in making our decisions. And while we might believe that a poker face is the ideal one to hold up to the world, is it possible that sometimes we're better off not hiding our emotions?

BET ON IT

Think about the commitment problem from an evolutionary point of view: maybe your genes have "programmed" you to occasionally sacrifice certain things now in order to better preserve and pass on your genes later.

You might suggest that the emotional component that sacrifices short-term gains for long-term stability is itself part of a larger, "rational" design that has been evolutionarily honed over thousands of generations. But you might not buy the remark about physical manifestations being part of the overall scheme.

The fact that our facial expressions and body language often gives us away is well-known. (There's even a TV show devoted to this idea.) Many people would wish for the ability to lie and cheat without their physiognomy letting them down. But Frank points out, for example, that the fact that someone is known to blush when lying would be to their advantage in situations requiring trust. (If they don't blush, they must be telling the truth. Others who don't blush might be lying.)

With someone who blushes, their signal is their word. You can probably think of other social interactions where such an unintentional enforcement of honesty leads to a positive outcome.

Altruism and Reciprocity

I'm sure you have read accounts of how soldiers have smothered a live grenade, thus giving their lives to save others, or about people who have jumped into icy water in an attempt to rescue complete strangers. These accounts are riveting and leave everyone, including scientists, wondering how such altruism can be explained.

Perhaps the best-known theory to explain such behavior is biologist Robert Trivers's concept called reciprocal altruism. Trivers's idea is that our emotions impel us to carry out altruistic acts. Emotions like guilt and sympathy can be the basis for altruistic behavior. But first, Trivers had to demonstrate that altruism really does exist, and he sought incontrovertible examples in nature.

Trivers's original example involves the symbiotic relationship between large "host" fish and small "cleaner" fish. The host fish have parasites that they cannot dislodge themselves. They allow the cleaner fish to live on their bodies and pick these parasites off. The host fish get cleaned and the cleaner fish get a meal.

This is not the end of the story. What prevents the host fish from simply gobbling up a cleaner fish when the task is done? This restraint is the core sign of altruism. Perhaps the host fish do not gobble up the cleaner fish because the cleaners are small and don't make up enough of a meal to outweigh the benefits of future cleaning. The fact that future interaction is delayed is another feature of true altruism, since the rewards for both are not immediate.

Complicating the issue is the presence of "cheater" fish. These resemble cleaners but instead of cleaning they swim up and take a hunk out of the host fish before scooting off. The cleaners make it easier for the host fish to find them again (and not get fooled by cheaters) by staying in the same place on the reef.

Of course, human interaction and altruism is more complicated than the cleaner fish example, but the main ideas when people are concerned are as follows:

- Reciprocal altruism is driven by the emotions.
- The reward for the altruistic act is not immediate.
- There must be a significant chance that the altruistic act will be rewarded, for example in a future interaction between the same people.
- Cheaters need to be recognized.

> **POISONED PAWN**
>
> Don't confuse reciprocal altruism with another type of altruism called *nonreciprocal altruism*. Nonreciprocal altruism is the idea that we perform altruistic acts for those who share our genes. We don't expect to be reciprocated for these behaviors. All other things being equal, the more closely we are related, the more likely the altruistic activity.

You might be fascinated by stories about cleaner fish and other creatures such as vampire bats that have been reported to engage in altruistic behavior. What exactly is the connection, though, with game theory?

The most frequently mentioned game theory model in the context of altruism is the Prisoner's Dilemma. With the PD, defection is a dominant strategy for both players, but the whole point of the PD is that mutual cooperation leaves them both better off.

Cooperating in the PD is an altruistic strategy. By cooperating, you leave yourself vulnerable to the sucker's payoff, but more importantly, you provide the possibility for the opponent to maximize his payoff when he defects.

In any one-and-done PD a cooperator can be taken advantage of. But now imagine a population of people going about their business and occasionally bumping into one another in a PD-style encounter.

Suppose you cooperated with Person A the first time and they defected. Here they are again. How will you behave this time? Many of us will want to pay them back by defecting. Doing so sends a strong signal that you are quick to punish someone who crosses you. But suppose Person A had cooperated last time? Given your own propensity to cooperate, and sensing a kindred spirit, you would have good reason to cooperate the second time around, too. This strategy, called "tit-for-tat," in which you are reiterating your opponent's previous move, seems to represent the essence of reciprocal altruism. You will learn more about the effectiveness of tit-for-tat in Chapter 22.

The Excitable Brain

Certain interactions with our fellow humans are enough to send us over the edge, like a bull seeing red. Road rage, for instance. But more generally, ever since the development of the functional MRI test for brain scanning about 20 years ago, we have gotten used to the idea that certain regions of our brains get "excited" by certain stimuli but not by others. A vast amount of research is ongoing to help us understand why we behave the way we do.

Emotion Trumps Rationality

For most of this book, I have worked with the principle that following the rules of rational decision theory sets the gold standard when it comes to making up our minds. Now I have come along and pointed out that, when the game is not as cut-and-dried as the ones in the textbooks, emotions can be a much more reliable guide to our actions than we might have believed.

Neuroscientist Antonio Damasio describes case histories of several patients that put to rest any doubt as to whether the emotions are a necessary part of decision-making, even "rational" decision-making. One patient (let's call him M) was a high-functioning man in his 30s who had developed a very large brain tumor that pressed on his frontal lobes. The tumor was successfully removed but some of the frontal lobe tissue, particularly on the right side, was damaged.

After surgery, M appeared to be the same intelligent, humorous, and charming person with full brain function. It didn't make sense that M suddenly could not hold one job after another, had gotten divorced (after having had a happy marriage before the tumor), got married and divorced again, and had made some terrible business, personal, and financial decisions.

He tested in the normal to superior range on an entire battery of psychological tests. But these tests failed to reveal that the part of M's brain that controlled his emotions was damaged by the tumor. The first clue to his emotional deficiency was a conversation in which M recounted the tragic events of his life in a severely cold-blooded and detached fashion.

The reason M was unable to make good decisions at his job or in his personal life was that the loss of his emotional makeup took away the ability to mediate and direct his rational decision-making apparatus. In reading a report at work, for example, he could spend an entire day on a single detail and still be unable to come to a decision about it. As Damasio explains, when we make a decision there is a path we must follow, a sequence of steps and choices that we must make, and lacking emotional functioning M was unable to stay on that path.

fMRI Studies, Choices, and Neuro-What?

In the past decade a huge amount of brain research has been going on and, in fact, has spawned some new fields of study. Just as we were getting used to hearing about neuroscience, we started to hear about a new field called *neuroeconomics*, and then another area called *neuromarketing*.

DEFINITION

Neuroeconomics is the scientific study of how economic decision-making is correlated with particular brain activity, while **neuromarketing** focuses on the brain's responses to stimuli such as advertisements and brand information.

Why do we even need to think about neuroscience in a game theory book? Now that we've begun to consider the emotional side of decision-making, we see that it's hard to isolate purely rational elements of decision problems such as which strategy to pursue in a particular game situation.

Sometimes, aspects of the situation that are not directly related to number crunching become salient. Once you know a certain strategy is called "cooperate," for example, you might generate a certain fondness or antipathy for this strategy that is unrelated to the payoff information.

One study that is already a neuromarketing classic involves Coke versus Pepsi. Taste tests between Coke and Pepsi have been going on for decades. But a study published by a Baylor University team in 2004 introduced a new element: fMRI results telling us what is going on in the brains of the test subjects.

The Baylor team conducted blind taste tests between Coke and Pepsi, and the results were that half of the drinkers preferred Pepsi. Moreover, their preferences were more pronounced on the scans. In other words, the brain activity in the Pepsi drinkers was higher than in the Coke drinkers. We can infer from these results that taste-wise, Pepsi would be less likely to lose customers.

But when the subjects were told which cola they were actually drinking, not only did three-quarters of them say they liked Coke better, but their brain activity showed remarkable changes. The information that they were drinking Coke lit up a different area of the brain than before, one associated with higher-level decision-making.

The Coke versus Pepsi example gives us a peek at the workings of neuromarketing. Marketers have known for a long time that our purchasing decisions are based on subjective, and not purely objective, criteria. But the hope for neuromarketing is that it will uncover the *neural substrate* for those subjective dimensions of our decision processes and, by doing so, will help pinpoint not only why we make the decisions we do, but how to influence those decisions.

DEFINITION

Given a specific behavior or psychological state, its **neural substrate** refers to the structures of the brain believed to be linked to the behavior or state.

Let's turn to neuroeconomics by introducing a well-known game called the Ultimatum game, which was first studied experimentally in 1982 by Werner Güth, Rolf Schmittberger, and Bernd Schwarze and has been a huge favorite of researchers ever since.

In the Ultimatum game there are two players. Each player understands that there is a certain sum of money, say $10, which is to be divided between the players only if they can agree on how it will be split. Player 1 will decide how much of the $10 to offer to Player 2. Player 2 receives the offer and either accepts it or not. If Player 2 rejects the offer neither player receives anything.

The obvious and "fair" solution is for Player 1 to offer $5 and for Player 2 to accept, thus splitting the money evenly. But the subgame perfect Nash equilibrium of the game is as follows: Player 1 offers Player 2 as small an amount as is feasible (say, $0.01) and Player 2 accepts (because $0.01 is better than nothing).

Later on I will dissect the Ultimatum game more thoroughly. But right now I want to try to make sense of the neural facet of our behavior in this situation. Suppose $10 were available between us and I offered you $1 of it. Would you accept or not? Most people would reject this offer even though it gives them a free dollar. When we hear the low offer, it starts a dialogue in our brains pitting our emotional side ("reject") against our cognitive side ("accept"). If you rejected my offer, what were you feeling?

Researchers put the Ultimatum game to good use in the days prior to fMRI testing. But being able to scan subjects' brains as they are reacting to offers in the game is even more exciting, and it goes an extra step toward understanding our emotional response to economic situations.

It turns out that people who reject the lowball offer experience a moral indignation, or disgust, that arises in a part of the brain called the anterior insula. However, the anterior insula is not the only part of the brain at work here. Imagine that I offered you $2 instead of $1, and suppose you are torn over how to respond.

There are now two parts of your brain that are dominating the back-and-forth argument you are having with yourself: the anterior insula, which wants you to tell me where to stick my $2, and the dorsolateral prefrontal cortex, which is trying to coax you into taking the money. All the while, another part of your brain—the anterior cingulate—is acting as the referee, weighing everything up. Fascinating stuff!

BET ON IT

If you knew that the low offers in an Ultimatum game were generated randomly by a computer, the element of disgust would vanish.

Before leaving the subject of neuroscience, let's look at one more example of the sort of work going on. You have seen a number of tricky games in this book: the Prisoner's Dilemma and Chicken, of course, and other games ranging from coordinating your actions to market games with asymmetric information. One emotion that is central to these interactions is trust.

If I'm weighing up how I think you will behave in a game of Chicken or a PD or a different encounter, either I will just play the "party line"—the rational, equilibrium strategy—or I might contemplate a different strategy if I have reason to believe that you will deviate from your rational strategy. Whether I am tempted to deviate will often depend on how much I trust you.

Not surprisingly, game theorists and neuroeconomists are interested in games of trust and what they bring out in our behavior. Interestingly, a lot of the recent work in this area revolves around the hormone oxytocin.

You may be aware that oxytocin plays a role in birth and breastfeeding. But oxytocin, among other things, appears to play a major role in our feelings of trust as well.

Remember the Centipede game from Chapter 4, where a pot of money is passed back and forth, and the pot doubles when a player refrains from taking the money? A number of researchers have designed similar trust games that work as follows. The first player receives an amount of money (say $10) and is given the option to transfer some of it to another, unseen player. Player 1 understands that whatever gets sent to Player 2 will be tripled. Player 2 is then given the sum of money that Player 1 has apportioned and, in turn, has the opportunity to return some of it to Player 1.

The first major finding in these trust games is that subjects, in general, do not play the (subgame perfect) Nash equilibrium strategy; they are more trusting than rational theory would indicate. But that's not the neuro-result. When the Player 2 subjects received money that they knew was intentionally sent, their feelings of being trusted caused an increase of oxytocin in their bloodstream. The amounts they sent back were highly correlated with the amounts they received.

In a different experiment with oxytocin and trust, subjects who were first administered oxytocin showed statistically more trust, as measured by the money they transferred, than those who received a placebo.

Neuroeconomics and neuromarketing are just getting off the ground, but they promise to uncover a great deal about how our emotional side contributes to our decision-making.

What Future Research Could Mean

When I first heard about oxytocin and trust I thought perhaps at the next G-8 summit the world superpowers might spike each others' drinks with oxytocin before going to the bargaining table.

More seriously, we should not expect the entire world of economics and game theory to be turned upside down simply because we learn what the neural correlates of certain behaviors are. But on the neuromarketing side, at the very least we can expect to gather more insight into what sorts of branding and advertising campaigns are most effective.

With respect to economics, and game theory in particular, neuroscience research will increase our understanding of what it means to be rational and irrational. The sophisticated models that economists and game theorists developed in the twentieth century were based on the idea of *homo economicus*—humans as purely rational and unemotional decision makers.

But over the past couple of decades a great deal of research seems to show that people often do not behave as the models indicate they should. Neuroscience might help us understand why certain stimuli or conditions cause us to act more rationally or more emotionally, and with this knowledge, we just might become successful at designing social and economic interactions that are more fruitful and less likely to break down. To put it differently, maybe someday our knowledge of how the brain works will help us dissolve personal, local, and global conflicts.

The Least You Need to Know

- Animal interactions feature some of the same games we play, and their behaviors often match what theory predicts.
- An evolutionarily stable strategy (ESS) in a population is one that cannot be successfully invaded by a different strategy.
- Our emotions enable us to commit to strategies that are costly in the short term but beneficial in the long term.
- Our emotional makeup is a necessary component of our rational decision-making ability.
- Neuroscientists are starting to identify the parts of the brain that are responsible for our rational as well as emotional decisions.

Aligning Theory with Behavior

Chapter 18

In This Chapter

- Quantifying our preferences
- Distinguishing theory from behavior
- Making sense of risk attitudes
- Understanding why losing hurts
- Misreading probabilities

I have frequently mentioned how the payoffs in games—which represent monetary or other outcomes—are often measured as utilities. This chapter further explores what utilities really measure and how they work.

This chapter also considers the ways people's appraisal of expected payoffs and their behavior in game situations either holds to or deviates from standard utility theory.

Utility Theory

It's time to get this nebulous idea of "utility" straight. The term itself was coined by scientist Daniel Bernoulli, who in 1738 proposed a solution to a curious expected value conundrum called the St. Petersburg Paradox.

Here's the paradox: suppose you and I play the following game. I flip a fair coin until the first heads appears. This first heads turns up on the nth flip and when it does, I will pay you 2^n dollars (= 2 * 2 * ... * 2, n times). But you first have to pay me a fee to play this game. How much are you willing to pay?

Think it over. I suspect you wouldn't pay a huge fortune to play this game. Now, using our expected value tools, let's compute the expected value of the game. The probability that the first flip is heads is $\frac{1}{2}$, and if this happens, I pay you $2^1 = 2$. With probability $(\frac{1}{2}) * (\frac{1}{2}) = \frac{1}{4}$, the first head comes on the second flip and I pay you $4. Get the pattern? The expected value of the game is then:

$$\frac{1}{2}*2 + \frac{1}{4}*4 + \frac{1}{8}*8 + \ldots = 1 + 1 + 1 + \cdots = \infty.$$

The paradox is that the expected value of the game is infinite, meaning that you should be prepared to pay a huge amount (say, your life savings) to play this game, but no one would.

Bernoulli realized the crux of the issue is that the value, or actual worth, of money to us does not vary proportionally with the monetary amount. For example, think of everything you could do with a gift of $1 million. Now imagine getting $10 million. The $10 million is worth a lot more to you than the $1 million, but is it worth 10 times as much? Not to most people. And then $100 million, in turn, is probably not 10 times as valuable to you as $10 million is.

Another way to see this is, going from $0 to $1 million is great, going from $1 to $2 million is still pretty darned good, but going from $91 to $92 million? Not such a big deal. This situation involves what economists call diminishing returns. Each additional million dollars is not quite worth what the previous million was to you. Realizing this, Bernoulli started to think about how he could transform the numbers in a way that was consistent with people's perceptions. We can represent his idea using the following simple graph.

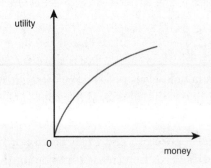

Utility for money exhibits diminishing returns

> **BET ON IT**
>
> Bernoulli used a logarithm function to represent his notion of utility for money, but since many people are allergic to logarithms, I use a square root function.

Bernoulli came up with a way to translate the monetary amounts into something he called utility in a way that represents the diminishing returns property. Here's an example of how this would work.

Let x equal an amount of money, in dollars. I denote the utility of x as $U(x) = 2\sqrt{x}$. So if you were given \$100, the utility would be measured as $2\sqrt{100} = 20$. If you were given \$1,000, the utility would be $2\sqrt{1000}$ which is about 63.2. This certainly satisfies the diminishing returns property.

If the square root function really does approximate someone's sense of utility, we can now figure out an answer to the St. Petersburg Paradox for that person. The expected utility of the gamble is as follows:

$$\frac{1}{2} * U(2) + \frac{1}{4} * U(4) + \frac{1}{8} * U(8) + \ldots$$

$$= \frac{1}{2} * 2\sqrt{2} + \frac{1}{4} * 2\sqrt{4} + \frac{1}{8} * 2\sqrt{8} + \ldots$$

Adding up this infinite series of terms requires us to resort to a couple of tricks that I will mercifully leave out; but when the math is done, the value of the gamble is \$5.83. In other words, someone whose utility function is represented by the square root function described above would be willing to pay at most \$5.83 to play the coin flipping game.

Now we know, more or less, what Bernoulli did. Two hundred years later, John von Neumann and Oskar Morgenstern revisited the same issue when writing their game theory book. They were able to nail down a precise and consistent approach that became the authoritative theory as far as economists were concerned.

Let's say that a player in a game needs to express his or her preferences over a set X of outcomes, where the outcomes are monetary or otherwise. Von Neumann and Morgenstern wanted to make sure that all outcomes in X could be ranked. So for any two outcomes A and B in X, either A is preferred to B or B is preferred to A (ties are allowed). This is called having a complete preference relation on the set X.

Next, this set of preferences must be transitive. (You saw this with voting.) So for any three alternatives A, B, and C, if A is preferred to B and B is preferred to C, then A must be preferred to C.

As you know, game analysis requires us to compare risky gambles, such as comparing one column of a game with another. Instead of using the casual word *gamble*, let's adopt the standard word *lottery*. A lottery over a set X of alternatives is just an assignment of probabilities to the alternatives so that the probabilities all add up to one. The St. Petersburg game is one example of a lottery.

Now think about a set X of outcomes and all the possible lotteries L that are possible for that set.

> **BET ON IT**
>
> Even a set of just two outcomes can have many lotteries defined on it. For example, suppose one outcome is $25,000 and the other outcome is $0. One lottery is 50-50 between the two outcomes; another lottery is 40-60; another lottery is 90-10; and so on.

Here are the von Neumann–Morgenstern conditions on what the preferences must obey. Let *pref* represent the preference relation: if A is preferred to B, you write A pref B.

- Pref is complete and transitive.
- Pref satisfies the "sure thing principle."

   The sure thing principle requires some explanation. Suppose you had a lottery G that you prefer to lottery F, and also that there is a lottery g that you prefer to lottery f. The sure thing principle says that for any weighted average between G and g, and the same weighted average between F and f, you would prefer the G/g scenario to the F/f scenario.

   Writing it mathematically, for any probability p, we have:

   $p * G + (1-p) * g$ pref $p * F + (1-p) * f$.

- Suppose there are three lotteries F, G, and H, with H pref G and G pref F. Then there is some way to average H and F so as to be indifferent to G. Writing it mathematically, you would say that there is some positive probability p such that: $p * H + (1-p) * F = G$ (i.e. indifferent).

   This final condition is called continuity or comparability.

To finish the job, all von Neumann and Morgenstern had to do is to quantify this preference relation. The final step is to set up a utility function U that measures preferences as numbers such that $U(A) \geq U(B)$ if and only if A is preferred to B.

Now that we have the definition of a utility function straight, let's see how to work with them.

The Standard Economic Perspective

Setting up the principles of utility theory is somewhat involved, but what utility theory says about our behavior is simple enough. The idea is this:

- When we make decisions, we are essentially looking at risky alternatives and therefore are dealing with lotteries.
- Our preferences over the risky alternatives can be expressed by a utility function as described by von Neumann and Morgenstern.
- We are expected utility maximizers. In other words, we go for the decision that maximizes our expected utility in the mathematical sense.

All the game theory calculations that we have made—calculating various equilibria, for example—are based on this concept. Let's see how the idea works in a particular risky situation.

Suppose you are on a game show. You can choose one of two doors. Behind one door is a briefcase with $25,000; behind the other door is a briefcase that's empty (that's right, with $0).

Instead of stressing about which door to pick, though, let me offer you an alternative. I'll offer you a sure payment of $10,000 instead. Just to make sure we have this right: either you choose the two-door gamble, which is a 50-50 chance at $25,000 or $0, or else you choose the sure $10,000. Which option do you prefer?

If we were using the expected value approach as we have been for most of this book, the answer is simple: the expected value of the gamble is 0.5 * ($25,000) + 0.5 * (0) = $12,500, which is greater than $10,000. A decision maker using this approach would choose the gamble, no problem.

But perhaps you find this a tough choice. If so, this literally means that the utility of the sure $10,000 is roughly equivalent to that of the 50-50 gamble. The $10,000 in this case is called a *certainty equivalent* for the corresponding lottery. Let's represent the situation with the following graph.

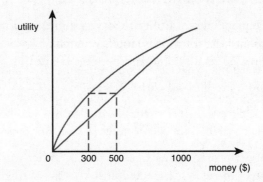

A lottery and its certainty equivalent

This person values a sure payment of $300 the same as a 50-50 lottery between $0 and $1,000.

DEFINITION

Given a particular utility function and a lottery, its **certainty equivalent** is the certain payoff amount that is judged to be equivalent in utility to the lottery.

The utility functions that economists and game theorists use are usually concave in shape, as in the accompanying figure. For any two points on the curve, when we draw a straight line between them, the curve is higher than the line. For one thing, concave functions satisfy the diminishing returns characteristic.

But the other point is this: the curve being higher than the line means that for any amount of money between the two points, your utility is higher when you get that amount for sure as opposed to a gamble between the two points with that expected value. For example, consider a 60-40 lottery between $25,000 and $0. The expected value is $15,000. The great majority of people—but not everybody—would prefer a sure $15,000 to this lottery. Such a preference means such people are *risk averse*.

DEFINITION

A decision maker is **risk averse** if they always prefer a certain payoff to a lottery that has an expected value that is equal to that payoff.

If your utility for money were strictly proportional—meaning that any monetary increase or decrease by an amount x would change your utility by exactly $U(x)$ no matter where your starting point was—then your utility function is linear; in other words, it can be plotted as a straight line.

So far, the idea of setting up utility functions and the way they seem to represent our preferences (at least for money) looks reasonable. Now we understand that when a player sees a payoff of $100 and another payoff of $200, the $200 is not necessarily twice as good to that player as the $100. This is a very important development. But, as we will now see, utility theory as it stands is not enough to explain a lot of our apparent preferences.

Some Confounding Paradoxes

There are a number of paradoxes, or contradictions, to utility theory that date back a half century or more. Let's look at the two best known of these older anomalies, after which we will see even more direct evidence that utility theory by itself misses a lot of what makes us tick.

In 1961 Daniel Ellsberg presented a curious property that is known as ambiguity aversion. Imagine Urn 1, filled with 50 red balls and 50 blue balls, and Urn 2, which also has 100 balls that are red or blue but in unknown proportions. You will select a random ball from the urn of your choice. Before you pick, you will place a bet on red or blue. If you are correct, you win $100, but if you are wrong, you win nothing.

NO BLUFFING

Daniel Ellsberg was a RAND Corporation analyst who also worked in the U.S. State Department. He is best known as the person who released the Pentagon Papers, which revealed some unfortunate facts about U.S. policy decisions in the Vietnam War.

Suppose you are planning to bet on red. Which urn will you choose?

Now let's say you are planning to bet on blue. Which urn will you choose now?

Most people will want Urn 1, with the known 50-50 mix, when betting on red. But then they'll want the same urn when they bet on blue! This is because the uncertainty regarding Urn 2 is unsettling to them. But let's look at the logic of this decision.

If, betting on red, you prefer Urn 1 to Urn 2, it appears that you believe the probability R_1 that you get a red ball in Urn 1 is higher than the probability R_2 that you get a red ball in Urn 2. In other words, your preference seems to imply $R_1 > R_2$.

But your unsettled feelings about Urn 2 also apply to the blue balls. If you prefer Urn 1 for the blue balls also, this would imply that you believe probability $B_1 > B_2$. But here's the contradiction: since $R_1 = 0.5$, then $R_2 < 0.5$. But then you would have to believe that $B_2 > 0.5$, because Urn 2 has 100 balls and you're indicating that less than half are red. But if $B_2 > 0.5$, you should have preferred Urn 2 to Urn 1—but you didn't.

Ellsberg's paradox might seem like a verbal sleight-of-hand. Maybe it isn't enough for you to want to give up on utility theory.

Prior to Ellsberg, Nobel Laureate Maurice Allais came up with a different paradox that directly contradicts the "sure thing principle." Allais constructed four different lotteries G, F, g, and f such that most people preferred G to F and also g to f. The sure thing principle says that a weighted average of G and g will have to be preferred to the same weighted average of F and f. But Allais found a counterexample to this; in other words, he showed that the F/f average had to beat the G/g average for the lotteries he constructed. In other words, people's actual preferences contradicted the theory.

At the time, despite these two inconsistencies (among others), most researchers were still pretty sold on utility theory. Their argument was that people's gut reactions to these hypothetical situations did not represent their actual preferences. Put differently, the argument was that utility theory was airtight but that people can get confused about what their true preferences are. Let's now delve into some much more convincing examples.

Prospect Theory

Prospect theory was launched by two psychologists, Nobel Laureate Daniel Kahneman and Amos Tversky, in a 1979 paper. Don't read too much into the name of the theory; *prospects* refer to different risky choices, as in a lottery, but mainly Kahneman and Tversky just liked the phrase.

NO BLUFFING

Although not an economist, Kahneman was awarded the Nobel Prize in Economic Sciences in 2002, six years after Tversky's untimely death. Tversky surely would have shared in this accolade.

Prospect theory, first of all, is based on how we actually behave, as opposed to utility theory, which is based on an elegant, but artificial, set of rules. There are four basic elements to prospect theory:

- Prospect theory extends utility theory into the realm of losses as well as gains.
- Prospect theory shows us that the way certain decisions are described, or "framed," has a huge impact on our thought processes.
- Related to the preceding point, prospect theory shows how our reference points influence how we look at changes in value.
- Prospect theory gives an account of our perceptions of probabilities.

Over the years, prospect theory has become very closely associated with two broader and somewhat overlapping fields called behavioral economics and behavioral decision theory. Regardless of the name, all of these subjects are concerned with the same basic question: *how do we make decisions, especially when gains, losses, probabilities, data, and time frames are involved?*

Gains vs. Losses

Let me ask you a simple question: suppose I flip a fair coin. If it comes up heads, I give you $50; if it's tails, you give me $50. Will you take this bet?

Next simple question: I flip the same coin, but this time I give you $51 if it's heads, you give me $50 if it's tails. Will you play this time?

Almost nobody will want to take the first gamble. Even though the expected value is zero (so it's a fair game), people just don't like the sound of it. Is that an indictment of utility theory? No, but it makes us pause a bit.

As you guessed, almost nobody would go for the second lottery either. Even though the expected value is slightly positive since the gain of $51 is bigger than the loss of $50, we still decline the opportunity. Now we suspect something is not quite right with utility theory.

Kahneman and Tversky explained that losses outweigh gains. The pain of losing $50 overshadows the joy of winning $50, which is why you don't take that bet.

Here's an exercise for you: assuming you still didn't take the bet when it was an even chance at winning $51 or losing $50, increase the $51 amount. Keep going until you find yourself feeling indifferent. Some people would still decline the bet when they could win as much as $100, while others might take the bet when the gain is just $60.

Realizing that losses pack more punch than gains is a revelation. But Kahneman and Tversky weren't finished with this concept yet. What they called the reflection effect is illustrated by the following situation.

Suppose you are offered the following choice: you can pick a lottery where you win $4,000 with 80 percent probability or $0 with 20 percent probability; or you can have $3,000 for sure. Notice that the expected value of the lottery is $3,200, which is more than $3,000. Eighty percent of the people Kahneman and Tversky polled chose the sure $3,000.

Now you are offered this choice: you can pick a lottery where you lose $4,000 with 80 percent probability or $0 with 20 percent probability; or you can pay $3,000 for sure. The expected value of the lottery now is -$3,200, but 92 percent of the experimental subjects went for the gamble instead of paying the sure $3,000.

The reflection effect shows that the great majority of us are risk-averse when contemplating gains, but are *risk-seeking* when facing losses.

 DEFINITION

A decision maker is **risk-seeking** if they always prefer a lottery to a certain payoff equal to the expected value of the lottery.

The following graph shows an *S*-shaped utility function that is typical for most people. In the positive half—the realm of gains—it is exactly the same as the one we've already seen. But in negative territory—the domain of losses—it illustrates how we are risk-seeking.

S-shaped utility function

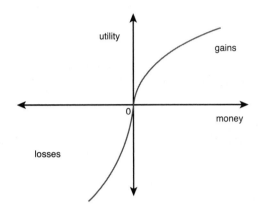

The curve is steeper for losses than for gains

Two things to notice about this utility function. It is steeper in negative territory than it is in positive territory. This demonstrates what Kahneman and Tversky called loss aversion: for the great majority of us, losses outweigh gains.

But also notice how, for any two points in the negative region, a straight line between the two points lies above the curve. This shows that we prefer a lottery between those two outcomes to the utility derived from the certain outcomes as plotted by the curve. Such a curve is called convex and it shows precisely the risk-seeking property.

 BET ON IT

Researchers have run prospect theory experiments all over the world using local currency and, when appropriate, converting it into sums with similar buying power as compared to the nation where the original data was collected. People all over the globe are remarkably consistent in their responses in these situations.

Practical-minded people might wonder what to make of all of this. Utility theory seemed a bit abstract, and while prospect theory promises to widen our understanding, some might wonder what it adds to our gaming savvy. Let me throw out a few tidbits here and in the next chapter to keep things practical.

For now, let me suggest how the government might increase tax revenue by appealing to our penchant for gambling. (By the way, I'm kidding, but not kidding.) Suppose you finish your tax return using a software package and when all is said and done you owe the government $3,000. Your software will offer you a couple of the usual options to pay this sum.

Here's one more option: the software allows you to pick a random number from 1 to 100—rolling the dice, so to speak. Eighty percent of the time you will end up having to pay $4,000, but 20 percent of the time your entire bill will be waived and you won't owe a penny. Let's suppose that the software is reliable and no one gets cheated on the $^{80}/_{20}$ randomization.

If people act in real life the way they do in experimental situations, most will go for the gamble. On the average, the government will rake in more than the $3,000 that they are owed; and while many people will have to cough up $4,000 instead of $3,000, at least they were given a chance to pretend they were in Vegas.

Relative Value

Another lesson of prospect theory is that when we consider changes in value, the starting point makes a difference, which is not the case in utility theory.

Suppose you are going to purchase a refrigerator. You can buy it at the nearby store for $995, but if you go across town you can get it for $990. Then you discover that your inkjet cartridge is running low. You can get one at the nearby office supply store for $20, but across town you can buy it for $15. Which deal is going to grab your attention?

In both cases you are saving $5. But most people aren't going to bother going out of their way to save $5 on a $1,000 fridge, while many people will go out of their way to save $5 on an inkjet cartridge, because they realize they are saving 25 percent.

Retailers understand this kind of behavior extremely well. The home improvement and big box stores know that you will not inconvenience yourself to save $5 on a refrigerator. They'll have to entice you with some other kind of deal. But an office supply store knows that you might show up to save $5 on some accessory.

There's another way to take advantage of this human tendency. Suppose you're on vacation. You've spent a fortune on your plane tickets and hotel. It's your only trip this year, and you're determined to make it special. After dinner you are wandering around and you find your way into a gift shop. They have some nice sunglasses for

$120. Ordinarily, you wouldn't pay $120 for a pair of sunglasses, but you've already spent a fortune getting to this place, so you say why not, and you buy them.

It all depends on your reference point. If you've spent thousands on your vacation, then $120 for sunglasses is inconsequential. But when you're shopping for an inkjet cartridge, your reference point is spending $20, so a drop to $15 is actually a big deal.

Risky Business

Kahneman and Tversky also discovered that when they would state probabilities to subjects in their experiments, the subjects seemed to make a mental adjustment or translation of these probabilities in some very interesting ways.

Their general finding is that most people underestimate most probabilities. For example, if I say to you that a certain event has a 40 percent chance of happening, when you hear me say "40 percent" the weight that you actually assign to the event mentally will probably be a smaller number.

However, Kahneman and Tversky also discovered that most of us overestimate very small probabilities. So if I tell you that a certain event has a probability of 1 in a 1,000, in your mind you are probably assigning a larger weight than 0.001.

To bring out this propensity, let's do a few examples. Suppose I offer you the following two choices with equal expected value:

a. A 40 percent chance at $5,000

b. An 80 percent chance at $2,500

Most people opt for the 80 percent chance at $2,500. If you mentally adjust both, then offer b retains more of its value.

What about very small probabilities? Consider the next pair of options:

c. A 0.001 chance at $5,000

d. A 0.002 chance at $2,500

Even though this pair offers the same expected value, most people prefer c. Evidently, in their minds they overweighted the 0.001 more than they overweighted the 0.002. The two probabilities are so small that people think they're almost the same, and then of course they go for the $5,000 prize.

Now, consider buying a lottery ticket. Suppose I offer you the following choice, again between alternatives of equal expected value:

e. A 0.001 chance at $5,000

f. A 100 percent chance at $5 (that's right, $5 for sure)

Most people prefer the lottery ticket–feel of choice e.

Finally, let's think insurance policy instead of lottery. Consider the following pair:

g. A 0.001 chance of losing $5,000

h. A 100 percent chance at losing $5 (a sure payment of $5)

Most people prefer h to g.

There are surely other reasons we buy lottery tickets and insurance policies, but prospect theory certainly seems to go part of the way toward explaining our attitudes toward risky alternatives.

The Least You Need to Know

- Utility functions quantify our preferences in accordance with a consistent theory.
- Experimental results appear to contradict utility theory.
- Prospect theory uses experimental data to explain our inconsistencies.
- Losses have more impact than gains.
- Our reference points influence changes in value.
- We tend to overestimate very small probabilities and underestimate other probabilities.

Behavioral Decision Theory

Chapter 19

In This Chapter

- How perspective affects choice
- How reference points influence choice
- How time affects choice
- How ownership affects choice

Chapter 18 started with utility theory but ended with a clinic on how prospect theory explains the quirky ways in which we perceive risky outcomes. In this chapter I continue with one more lesson of prospect theory and then branch out to explore some related topics in behavioral economics.

The main lessons of this chapter are how identical decisions can appear very different to us depending on their storylines, their contexts, and their timeframes. These lessons heighten our self-awareness, help us understand how others think, and therefore improve our ability to strategize in game situations. As I will show you, rather than discovering that we are "irrational," we are finding that our deviations from mathematical theory are fairly predictable. To put it differently, there is a method to our madness.

Frames of Reference and Mental Accounting

Much of prospect theory and behavioral economics explores our risk attitudes toward monetary decisions. Psychologists Daniel Kahneman and Amos Tversky—whose research provided the basis for Chapter 18—also studied our attitudes toward other types of outcomes, including those involving human lives. Here's another experiment of theirs.

Suppose you are a public health official. There is an outbreak of a deadly disease in your community and 600 people are expected to die. You have two different options to employ. Option A will save 200 people. But with option B, $1/3$ of the time all 600 people will be saved and $2/3$ of the time no one will be saved. Which option will you choose?

Now let's consider another situation. You're still in public health. There is another, equally deadly outbreak of disease in your community, meaning that again, 600 people are expected to die. You have two options to consider. With option C, 400 of the 600 people will die. With option D, $1/3$ of the time no one will die and $2/3$ of the time all 600 people will die. Which option will you choose this time?

In the first situation, 72 percent of the experimental subjects chose option A, where 200 people are saved for sure. The second situation is in fact identical to the first one, but here, 78 percent of the experimental subjects chose the $1/3$, $2/3$ gamble presented in option D. Same situation, but the different "framing" of the problem led to a very different decision.

The different behaviors can be explained as follows. We know that most people are risk-averse when looking at gains. In this setting, every life saved is something gained. We can therefore understand (when the situation is framed as saving lives) that most people go for the sure thing as opposed to the lottery with the same expected value.

However, the second situation poses the outcomes as loss of life. Most decision makers are risk-seeking when facing losses, so we can now see why most experimental subjects chose the lottery rather than the sure loss of life.

BET ON IT

Something as simple as describing a situation in terms of losses instead of gains is enough to turn us from a tendency to avoid risks to a tendency to take risks.

Let's draw back from life and death for a moment. Have you ever noticed how you are willing to purchase something in one circumstance but not another? Another curiosity uncovered by Kahneman and Tversky is a way we justify certain actions by means of what they called "mental accounting."

Suppose you bought a ticket to a play for $40. When you get to the theater you realize that you've lost the ticket. Are you willing to spend $40 for another ticket?

Now imagine that you are on your way to the theater, intending to buy the ticket there. When you go up to the ticket window you open your wallet and discover that you have lost $40 in cash. Are you still going to buy the ticket?

In the original experiment, more than half the people who lost the ticket would not buy another one. But 88 percent of those who lost the cash were willing to buy a ticket. The outcomes are the same in terms of monetary value but the subjects' responses are markedly different. How can this behavior be explained?

The mental accounting that Kahneman and Tversky describe means that we seem to assign our money to different funds for different activities. When we lose the ticket, it is as if the theater fund has been emptied; spending another $40 seems to go beyond the budget we'd allocated. But when we lose the cash instead, in our minds the loss comes out of some larger, pooled fund, and the theater fund is still intact.

BET ON IT

The framing effect as well as our propensity to use mental accounting are examples of what Kahneman and Tversky call *failure of invariance*. Many of us pride ourselves on our ability to think logically and consistently, but given all of the examples we have racked up, it is clear that human preferences do not follow the laws of expected utility theory.

Anchoring and Relativity

Very often the decisions we make involve comparisons to some other type of benchmark information. When those standards of comparison change significantly, watch out—our decisions might change as well.

Suppose we get 100 test subjects and randomly split them into two groups. We take one group aside and privately ask each person the following two questions:

a. Is the Columbia River longer or shorter than 1,500 miles?

b. How long do you think the Columbia River is?

With the second group, we privately ask each person:

c. Is the Columbia River longer or shorter than 500 miles?

d. How long do you think the Columbia River is?

I assume very few people know how long the Columbia River is. But giving the first group an *anchor* of 1,500 miles provides a point of comparison that is much longer than the anchor of 500 miles for the second group. When we average the responses to questions b and d, we will find that the first group believes, on average, that the Columbia River is much longer than the second group does.

DEFINITION

An **anchor** is a prior, numerical point of comparison.

Sometimes we engage in anchoring to irrelevant stimuli. Behavioral economist Dan Ariely describes one of his experiments in which subjects were privately bidding for various products like cordless keyboards, chocolates, and bottles of wine, but unknowingly were manipulated using a process that psychologists call priming.

The subjects were instructed to write down the highest amounts that they were willing to pay for the various items. This sounds straightforward enough, but prior to making the bids, the subjects had been instructed to write down the last two digits of their social security numbers.

Of course, merely writing down the last two digits of your Social Security number shouldn't have anything to do with how much you are willing to pay for a package of Neuhaus chocolates. And after the experiment was conducted, the subjects themselves reported no such influence. But the results proved otherwise. Overall, the subjects writing down digits from 80 to 99 bid 216 to 346 percent higher on average than subjects who wrote down numbers from 1 to 20.

The anchoring phenomenon sounds interesting but may not strike you as immediately relevant for game theory. One direct application is in the realm of two-person bargaining. Good bargainers try to establish a status quo, or make an initial demand, that is greatly tilted in their favor. If the adversary concedes this first step, then the eventual outcome is likely to favor that person who was able to legitimize their exaggerated demand.

There are a number of more subtle consequences of anchoring that are especially useful in marketing and pricing applications. As Ariely points out, restaurants often manipulate us in their menu pricing.

Suppose you are looking at a menu and there's an enticing entrée listed for $35. If that dish were the most expensive one on the menu, it might look too pricey and you reluctantly order something less expensive. To induce you to order this item, however, the menu might include a super-deluxe dish for $40. The $40 item is a "decoy" designed to lessen the impact of the other expensive dishes.

Now suppose you are in an electronics store and there is a newfangled gizmo that you find pretty fascinating. You look at the price tag of this item and it sells for $250. Is that a good deal? Well, $250 is a lot of money and you have nothing to compare it to, so with some serious regret you shuffle out of the store.

The company that developed this ingenious gizmo isn't likely to sell a lot of them, are they? But now they get smart. They develop a fancier version—with a couple more features—and put that on the shelf also. It's priced at $325. Now shoppers have something to compare. Maybe the company gets lucky and sells some of the $325 products, but more importantly the $250 gadget will look pretty good by comparison, and now customers will be a lot more likely to purchase it.

We've just seen how our decision-making can be altered when alternatives are added to the scene. Even doctors' judgments can be swayed in this fashion. In a study published in the *Journal of the American Medical Association*, family physicians were presented with a case of an older male with chronic hip pain. When doctors were given two choices—either to refer to an orthopedic specialist or to prescribe a particular medication—53 percent of them opted for referral.

The other family physicians in the study were given three choices. They could refer the patient to a specialist, prescribe the same medication as the other group, or prescribe a different medication. In this group of doctors, 72 percent opted for referral.

This physician study may remind you of the concept of irrelevant alternatives that we have seen earlier. If the third alternative (the other medication) was a relevant alternative, then presumably some doctors would have selected it, and the number of doctors opting to refer the patient would have decreased or stayed the same. And if the third option was truly irrelevant, then the number of physicians opting for referral should have stayed constant.

The doctors' choices not only defy independence of irrelevant alternatives, but seem to contradict our intuition from the retail example, too. In the retail example, adding a second gadget made the first one more attractive, but with the physicians, adding a second medication made the first medication less attractive. (Remember, more docs opted for referral.) How can this be explained?

In the retail example, the second gadget acted as a decoy, designed, by way of comparison, to make the first gadget look good. But with the physicians, the second medication was not a decoy. This additional option only served to confuse the doctors. Unable to decide between the two medications, they opted out of the decision by electing to refer the patient to a specialist instead.

> **POISONED PAWN**
>
> We can't simply assume that adding alternatives always increases or decreases our likelihood to select an option from the original set of choices. We need to understand whether the new alternative is either a truly irrelevant one, a decoy that will increase someone's interest in one of the established choices, or else is simply going to cause confusion.

The anchoring behavior we just reviewed is similar to what we saw with framing. With framing, the glass being half-empty or half-full, so to speak, makes a huge difference to us. With anchoring, the context of the situation—in these cases, the relative comparisons we carry out—makes all the difference in the world.

Intertemporal Choice

Suppose I offered you the option of receiving $15 now or else receiving a different sum a month from now. When behavioral economist Richard Thaler asked subjects how much they would need to receive a month from now in order to match the immediate payment of $15, he found that the median response was $20 (in other words, half of all respondents answered less than $20, and half answered at least $20).

This may seem reasonable to you—if you had to wait a month to get the money, you'd find it natural to ask for more. But now think about this: suppose I offered you either $15 to be received a year from now or $20 in 13 months. You would probably prefer the $20 in 13 months. In fact, most people who prefer $15 now to $20 in a month's time will find that they prefer $20 in 13 months to $15 in a year.

Two interesting things are going on here. First of all, when people have to wait a long time to get paid, they don't mind too much if they have to wait a little longer. So most people, given that they have to wait a year to get $15, would happily wait that extra month to get $20 instead.

Now notice the reversal of preferences: the same people who say they prefer $20 in 13 months to $15 in 12 months will turn around after the 12 months pass and will realize they'd rather have the $15 right away than wait another month for the extra $5!

The reason that behavioral economists try to discover our preference patterns for money over time is that so many of life's decisions take place over extended periods.

This is true in game situations as well—for example, I brought up the importance of the time horizon when we looked at the Tragedy of the Commons. Studying how our preferences for money (and other things) change over time is a fascinating area of study called intertemporal choice.

One important human characteristic that has been supported by most studies in this area is that as the time horizon lengthens, people's *discount rate* decreases. Let's run some numbers to understand what this means:

DEFINITION

The **discount rate** is an annual measure of how the value of money changes over time, usually due to interest or inflation. For our purposes it is the same as an annual interest rate or rate of inflation.

- In Thaler's study, $15 today was worth the same to people (as measured by the median response) as $20 a month from now. This 33 percent increase in one month works out to an interest rate (compounded daily) of 354 percent annually.

- The $15 today was worth the same to the participants (again, median response) as $50 after one year. Going from $15 to $50 in one year works out to an interest rate (compounded daily) of 121 percent annually.

- Finally, the $15 today was worth the same to the participants (median response) as $100 in 10 years. Going from $15 to $100 in 10 years works out to an interest rate (compounded daily) of about 19 percent annually.

As you can see, the longer the time horizon, the less the discounting. One way to think of this is that people are very impatient over short periods of time, and this impatience is reflected by their charging a very high rate of interest in their minds. As the time periods get longer, people seem to adopt a more relaxed mind-set. The economists who study this phenomenon call it *hyperbolic discounting* (for mathematical reasons we will skip here).

Thaler turned up some other interesting patterns. The preferences between $15 now and $20 in a month did not scale up proportionally to large sums of money. He had subjects consider a $3,000 prize to be received immediately, and asked them how much they would need to receive in one month to match the $3,000. The median response was $3,100. Going from 15 to 20 is an increase of 33 percent, but going from $3,000 to $3,100 is an increase of only 3.3 percent.

One of the important findings of prospect theory is that people treat losses differently from gains. Thaler investigated this behavior in the current setting as well. He had subjects consider fines they would have to pay. Subjects were asked to consider a $100 fine to be paid immediately, and then to consider paying the fine in a year's time. What payment, after a year's delay, would they feel was equal to the immediate fine of $100? It turned out that the median response here was $118.

When considering gains and losses over time, once again our behavior differs between the two. If we have to wait for a prize, the amount has to appreciate considerably for it not to lose value for us; but delaying the payment of a fine does not exhibit the same pattern. Most people, if hit with a $100 fine today but given the opportunity to pay $125 a year from now, will not put it off.

By the way, lots of retailers, for example, furniture stores, occasionally offer deals where you pay nothing for (say) 12 months. These deals may entice a lot of customers. But no retailer says, pay nothing now but you'll owe 25 percent more a year from now. If they did, there would presumably be few customers who would jump at the chance.

What have we learned so far about how people's preferences for money change over time?

- We have a powerful urge to receive money now as opposed to waiting a month or more.
- This urge subsides as the payments increase.
- When our time horizon lengthens, we become more patient about receiving payment. (Going from a 0- to a 1-month wait is a big deal, but going from a 12-month wait to a 13-month wait is not.)
- We are not willing to pay much of a premium to be able to put off an immediate payment.

There is a huge body of research in the area of intertemporal choice, but the sampling I have given you provides the results that are most relevant to people's behavior in games. Remember, our focus is on payoffs and how we and others interpret those payoffs.

Prospect theory gives us some deep insights into our attitudes toward risky alternatives; intertemporal choice helps us understand our preferences for outcomes over both shorter and longer periods of time.

The Endowment Effect

Often in our strategic interactions, a player is reluctant to give up the status quo in favor of a promising alternative. The expression, "a bird in the hand is worth two in the bush," sums up our preferences for sure things as opposed to risky alternatives.

BET ON IT

Sometimes the reluctance to leave the status quo is just that: we're afraid of change because the downside of the alternative outweighs the upside. This reluctance is called *status quo bias*.

But sometimes the reluctance to give up the status quo is actually a reluctance to part with something we own. Richard Thaler coined the expression "endowment effect" to represent our tendency to value things more once we get attached to them.

In one study, Cornell University undergraduates were randomly split into two groups. Half of the students were given university mugs, which sold for $6 at the school bookstore. The other half of the students were not given mugs. When a market was set up to buy and sell the mugs among the students, the experimenters discovered that the owners valued the mugs at a median price of $5.25, while the buyers only valued the mugs in the $2.25 to $2.75 range.

A different study at Duke University examined students who desired tickets to the NCAA Final Four basketball tournament. Students were first required to wait around in tents for days; those who stuck it out then had their names submitted in a lottery to actually award the tickets. Given that all the students in the lottery pool were rabid basketball fans, the question was, would those who randomly won the tickets in the lottery value them any differently from those who did not win?

Out of over 100 students who were queried, the average price that students without tickets were willing to pay was about $170. But astonishingly, those who had been awarded the tickets asked for about $2,400 to sell them. Clearly, the outcome of being awarded a ticket transformed its value in the minds of the lucky owners.

The endowment effect is another important psychological behavior that we have to be aware of when we study strategic interaction. People who are interested in behavioral finance will observe this effect in the stock market.

Suppose you bought some shares of a stock that you had good reason to believe was destined to appreciate in value. Months later, however, the stock price has continued to go down and at this point you don't see any particular reason to be optimistic about a turnaround. Your rational side wants to sell the stock and cut any further losses, but you just can't bring yourself to do it.

Why can't we pull the trigger more easily and get rid of that disappointing stock? One reason is loss aversion. We hate to lose, and actually selling the stock is an admission of defeat. If we hold the stock longer, we haven't officially lost yet.

But there's also the endowment effect. Suppose the price is currently $28 a share. You would certainly not purchase this stock at this price—indeed, you're thinking about getting rid of it—but something about owning it makes you think the stock is worth more than it objectively is.

As this chapter comes to a close, let me emphasize the following in summing up these lessons on utility theory, prospect theory, and behavioral economics: the wealth of quirky examples of our "irrational" decision-making does not mean that we are entirely irrational and that our decision-making is chaotic.

Rather, the evidence is that we are rational to some extent, and when our rationality is bounded, the ways in which we deviate from cold-blooded mathematical computation are not only understandable, but fairly systematic. In other words, we are fairly consistent in our inconsistency—or, as I've already put it, there is a method to our madness!

The Least You Need to Know

- When an identical problem is expressed with outcomes described as losses instead of gains, the decisions we make can be entirely reversed.
- Our perceptions of value can be markedly influenced by comparisons that we take to be irrelevant and harmless.
- We are impatient when imminent rewards are delayed but not as upset when rewards expected far into the future are delayed.
- Once we own something, its value to us seems to increase.

Strategic Behavior in the Lab

Chapter 20

In This Chapter

- Behavioral game theory
- Measuring trust
- Self-interest versus altruism
- Getting on the same page

As I've tried to make clear, game theory first and foremost is an attempt to use mathematics to model strategic interaction. In particular, it is a theory that seeks to find the best strategies for players to adopt, given that they are ultimately concerned with their expected payoffs.

But as I've also emphasized, our very human urges and preferences often contradict the choices that an impartial mathematical theory would prescribe. For this reason, if game theory is to stand up as a social as well as a mathematical science, we need to understand when the human element converges to, and diverges from, the mathematical. In this chapter I report on some fascinating results from people's strategizing in game situations.

Dictators and Trust in the Laboratory

If all players in the world were robots programmed to play Nash equilibrium strategies, then the study of game theory could bypass any need to discuss concepts like utility or report on emotional or "irrational" play as I have done with prospect theory, neuroeconomics, and behavioral economics in general.

But, since we are human, any study of strategic interaction needs to include behavioral findings. Psychological motivators like trust may play no role in Nash equilibrium analysis, but they are central to situations where collective payoffs may be improved through nonequilibrium play.

Just as sciences like physics have both a theoretical and experimental side—where the experiments show whether the theory passes muster—game theory, too, is increasingly becoming a field with both theory and practice. Behavioral game theory is game theory's experimental side, where subjects are thrust into game situations, often using real money, and their actions are observed and analyzed.

As I mentioned in Chapter 17, trust is a key element in a number of game situations. In particular, trust is an important factor for players deciding whether to deviate from a coldly calculated equilibrium strategy. In the following sections I examine a couple of different behavioral games that focus on how much people can be trusted to put aside their self-interest for the good of others.

The Dictator Game

The most basic game of trust that game theorists have studied is called the Dictator game. Here's how it works: there are two players and a sum of money, say $10, to be divided between them. Player 1 is completely free to determine how much of the $10 to keep and how much to give to Player 2. Player 2 performs no action in the game, being merely in the background (and perhaps in Player 1's conscience). When people are in this situation, how do they actually behave?

In the first such study (1986), the amount to be shared was $20 and the subjects had two specific choices: either to give half, or $10, to Player 2, or to give Player 2 $2, while keeping $18. About 75 percent of the people opted for the 50-50 split, which is a pretty encouraging report on human nature. Alas, the studies that followed were nowhere near as heartening.

In a compilation of numerous studies, the average amount given to the other player ranged from 10 to 38 percent, with most results in the 20 percent neighborhood. Perhaps more telling, the proportion of people that kept all of the money was most frequently in the 30 to 40 percent range.

One study set out to elicit generosity from the players. In one group the Player 1's got to know who the other player was; in another group there was mutual identification; and in a third group there was communication between the players on top of that. In the mutual identification groups, the mean amount left for the other players was about half of the total, indicating that when players know one another, they tend to be more generous.

Another study, co-authored by Nobel Laureate Vernon Smith, was particularly ingenious and striking in its results. In this study subjects were shown into a private booth. The experiment was set up in a "double-blind" fashion so that the experimenters would not know how the subjects behaved individually. The subjects were given 10 $1 bills, 10 blank dollar-size slips of paper and an unmarked envelope. Each subject was instructed to leave the envelope with 10 pieces of paper—any mix of dollar bills or blank slips they wanted—in the envelope for the other person. The subjects could make off with any number of the dollar bills and no one would know.

The experimenters put the unmarked envelopes in a box to be examined later. The idea was that since the subjects' behavior, whether selfish or generous, could never be known, the experiment would elicit people's true colors, so to speak.

And did it ever! Over 60 percent of the subjects took all $10, leaving just blank slips for the unfortunate Player 2. And those who left money didn't leave much, mostly just $1 or $2.

Overall, the results in the Dictator game literature indicate that when players were identified by other players or by the experimenters, a significant number came under the spell of conscience and guilt. Maybe some people hoped to impress the experimenters with their generosity. But when players' actions were untraceable, only a small minority was compelled to share the money in anything like a fair manner.

While these outcomes don't fill me with pride to be a member of the human species, we should not take them as the final word on trust. Most of these players could not be trusted to part with money unilaterally in a one-and-done game, but what if the situation involved back-and-forth exchanges of money instead?

Trust in Stages

One interesting experiment involved two players making sequential decisions about how much money to keep and how much to give to the other player. Player 1 is given a sum of money M, and has to decide how much to keep (= K) and how much to give away. When Player 1 gives away $G = M - K$, he is aware that the money will grow at rate r. In other words, the amount G grows to $G * (1 + r)$. Player 2 receives this sum $G * (1 + r)$. She has to decide how much to keep and how much to return to Player 1.

As behavioral economist Colin Camerer explains, this game measures trust on the part of Player 1 (how much he is willing to risk Player 2's not reciprocating his good deed), and measures trustworthiness on the part of Player 2 (how much she is willing to return, given that she knows her sum was provided, after all, by Player 1).

In the initial experiment, there were 32 pairs of players, who happened to be students in Minnesota. The test was run in a "double blind" fashion, so the subjects did not get to know or communicate with one another. The initial sum M was $10 and the rate of growth r was 2, meaning that the money would triple. To make sure we understand how this works, here's one possible outcome:

1. Player 1 is given $10. He keeps $5 and invests $5.
2. The $5 grows to 5 * (1 + 2) = 5 * 3 = $15.
3. Player 2 receives $15. She keeps $10 and returns $5.
4. Player 1 and Player 2 each end up with $10.

Notice that these hypothetical strategies I've chosen do not maximize the payoffs for the players. If Player 1 completely trusted Player 2, he would keep nothing at first and invest all $10. The $10 would grow to 10 * (1 + 2) = $30, at which point it is up to Player 2 to decide how much to return.

Out of the 32 pairs, 5 of the Player 1's invested all $10. Three of those five got at least $15 back, but the other two got just $1 and $0. Overall, the average amount invested was about $5. In other words, on average, Player 1's kept $5 of the allocated $10. And, on average, Player 2's returned a little less than they received. Furthermore, there was basically no relationship between the amount of trust on the part of Player 1 and the amount of trustworthiness displayed by Player 2.

> **NO BLUFFING**
>
> Before you conclude anything about Minnesotans, I have to point out that this type of experiment has been replicated in a number of places around the world, and, while there are some interesting regional or national variations, the Minnesota results are fairly typical.

This back-and-forth exchange may remind you of the Centipede game that we looked at in Chapter 4. Remember how it goes: there are two players and an initial pot of money. Each player has two strategies available, "take" or "pass." Player 1 can choose to take a certain proportion of the initial pot, leaving the rest for Player 2. Alternatively, Player 1 can pass. If he passes, the pot doubles.

Now it is Player 2's turn. She can take that same percentage of the bigger pot, leaving the rest for Player 1, or she can pass, in which case the pot doubles again. The players alternate, and whenever a player decides to take, the two shares are given out and the game ends. The game is played at most a certain predetermined number of rounds. If all plays are "pass," then the amount in the pot after the last play is distributed according to the proportions favoring the next player.

In one behavioral study of the Centipede game, the two players started out with $0.50 total. The take percentage was 80 percent, and the game ended after four rounds. The following diagram shows the payoffs to the two players after each decision.

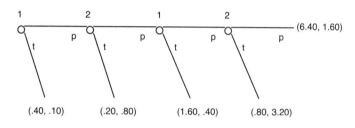

Four-Round Centipede Game

t = take
p = pass

Payoffs in dollars

In Chapter 4 I introduced the idea of subgame perfect equilibrium by imagining Player 2, for instance, trying to persuade Player 1 to pass by promising to pass himself when it is his turn. This gambit, as we saw, is unconvincing. Let's go the final play of the game and do an analysis, called *backward induction*, going back in time all the way to the first move.

DEFINITION

Backward induction is the process of finding out the best strategy to play at the end of the game and then working backward to find the best strategy for the second-to-last play and so on all the way back to the beginning.

Let's look at the final decision, which is made by Player 2 at the far right of the diagram. The pot has reached $4. If Player 2 takes, she ends up with 80 percent of $4, which is $3.20. If she passes, then the pot doubles to $8 and the game ends with 80 percent of the $8 going to Player 1 and the remaining 20 percent, or $1.60, going to Player 2. Since $3.20 from taking is better for Player 2 than the $1.60, a rational Player 2 will take on the last play.

Anticipating this, what will Player 1 do on the play before? He realizes that Player 2 will take on the last play, leaving him with only $0.80. If Player 1 pre-empts this outcome by taking on the second-to-last play, he will get $1.60, which is better than the $0.80 he would be left with if he passed. So a rational Player 1 will take on the second-to-last play.

If we extend this reasoning backward in time, we find that the rational strategy for Player 2 at the second move of the game is to take, since taking leaves her with $0.80, which is better than the $0.40 she is left with if Player 1 takes on the next move. Similarly, the best strategy for Player 1 at the very first move of the game is to take as well.

So by the logic of backward induction, rational players will take at every move of this game. In particular, this means that with rational play, Player 1 will take right away and the game will end with Player 1 getting $0.40 and Player 2 getting $0.10.

This grim view of the interaction makes the Centipede game sound like a Prisoner's Dilemma. We might start to lament: if only players trusted each other a little more! Let's see what happens in real life, or at least, in a real-life laboratory setting.

In real life, players do not play their subgame perfect Nash equilibrium strategies in the Centipede game. In this study fewer than 10 percent of the players played take at the first move. As the game went on, the proportion of players playing take increased at each move. This makes sense. However, the really surprising result was that around 20 percent of the players did not play take at the last move of the game. These players passed, settling for a payoff of $1.60 rather than enjoying a $3.20 payoff by taking.

What can we make of these results? Even though the cold rationality of backward induction tells us that taking is always best, clearly the experimental subjects felt it was more rational to pass and thereby increase the size of the pot, at least in the early going.

POISONED PAWN

Backward induction is a mathematical argument that is frequently used in game theory. While it has an unassailable logic, often the implications are unsettling. In this case the "rational" strategy is to take immediately, generating a payoff of $0.40 for Player 1. Imagine if the game had 10 (not to mention 100) moves. You may find it much "more rational" for the players to play pass for several rounds to enlarge the pot and then strategize in the endgame stages. The point is that these induction arguments stretch out over many plays and can ignore some very important features of the game.

What about those 20 or so percent of players who passed on the last play, thereby cutting their own payoffs from $3.20 to $1.60? Don't forget that we are talking about trust—and trustworthiness—here, and one explanation is that these Player 2's knew full well that had the Player 1's taken on the previous move, the 2's would have been left with only $0.40. So 20 percent of the Player 2's were showing some gratitude here. Perhaps they derived significant utility from their good gesture, enough to outweigh the small opportunity cost.

Behavior in the Ultimatum Game

I introduced the Ultimatum game to you in Chapter 17. Remember how it works: there are two players and a given sum of money, say $10, to be shared, provided the players agree on the outcome. Player 1 offers some of the $10 to Player 2. Player 2 either accepts, in which case the money is parceled out and the game ends; or Player 2 rejects the offer, and the game ends with neither player getting anything. We learned that when people feel insulted by a low offer, their disgust is clearly registered by fMRI scans.

Note that the Dictator game is a one-sided Ultimatum game in which Player 2 does not get the opportunity to play. Behaviorally, we should expect a big difference between these two games. In the Dictator game, Player 2 is powerless and Player 1 acts according to some mix of greed and generosity. But in the Ultimatum game, Player 1 has to act strategically. He realizes that Player 2 might reject a low offer, so besides considering whatever payment he might impulsively offer, he has to size up how he believes Player 2 will react.

Another reminder is that the subgame perfect Nash equilibrium of the game is for Player 1 to maximize his share by offering as small an amount as is possible (say, $0.01) to Player 2, and for Player 2 to accept this offer (since $0.01 is better than nothing).

There is a huge literature on the Ultimatum game and it is not my objective to discuss every wrinkle of what researchers have found. However, I will survey the more interesting and important results. Before we get into some of those, though, we should step back and not only grasp what the game is measuring but also show that the game is more than an abstraction and distillation of some basic human tendencies.

Behavioral economist Colin Camerer makes a remarkable historical point that is worth retelling since it highlights how the thought process in the Ultimatum game underlies many practical decisions that, for example, legislators or business people might face.

At the 1787 Constitutional Convention of the fledgling United States, one topic of discussion was how to incorporate new states that would surely join the Union as the nation expanded westward. One option was that the original 13 states would retain a pre-eminent role with the newer states adopting a second-tier status. But George Mason argued that if the western states were offered a lesser form of statehood, they might not join the Union or would otherwise plan to secede. Mason's argument held the day and, as we know, all states were given equal voting power.

BET ON IT

The Convention delegates were, in essence, anticipating an Ultimatum game. They understood the risk of offering less than a fair share. In the Ultimatum game proper, some money is at stake and that's it; but in real life, when people are given deals that are less than fair, they tend to have long and bitter memories that may come back to haunt the instigator someday.

In Ultimatum game (and other) experiments, we can't wait around for umpteen years to see if someone seeks revenge from having been mistreated ages ago. But we can obtain some clues regarding, for example, what most people are inclined to offer, and how far recipients will bend before they break—in other words, how low is too low. We will also discover whether other considerations influence both the size of the offers and the propensity to accept them.

Baseline Results

Out of many studies that have been compiled, let me summarize some overall findings:

- The average amount offered, the most frequent offer, and the median amount offered, are typically in the neighborhood of 40 percent.

- These typical offers—in the 35 to 50 percent range—are accepted the great majority of the time.

- Offers in the 20 percent range are rejected about half the time.

- There is some variation across the globe, which seems to indicate various cultural influences.

Note that behavior in the Ultimatum game is significantly different from that in the Dictator game. Ultimatum game offers are higher, indicating that Player 1's (making the offer) are wary that Player 2's might reject low offers. And Player 1's have good reason to be wary, since, for example, a 20 percent offer has a high chance of being rejected.

A key question is whether Player 1's, in general, tried to maximize his take given how he expected Player 2's to respond. For example, suppose someone in a Player 1 role was aware that a 20 percent offer would be rejected half the time—and was therefore too risky to try—but thought they could almost surely get away with a 30 percent offer. Do such players offer 30 percent, which is the minimum they think they can get away with, or are they typically more generous than that? The results show that players, on average, are somewhat more generous than they need to be.

One way to sum up the results is to say that in the Ultimatum game, people do play strategically but they also have something of an altruistic streak.

More Penetrating Findings

One effect that was studied in a number of experiments was whether repeating the game, thereby building up subjects' experience, had any effect on behavior. Some studies reported that offers tended to decline somewhat over time, as did rejection percentages. This would indicate that Player 2's gradually got used to lower offers and, as a result, punished Player 1's less of the time. Other studies, however, found very little such effect over time.

Further studies have looked into whether raising the stakes would make a difference. Suppose we had $10 available and I offered you 10 percent, or $1, of it. You'll probably reject my low offer. But what if there were $1 million available and I offered 10 percent, or $100,000, to you? You might generate the same anger, indignation, and disgust as with the $1, but I bet you would nevertheless accept the $100,000. As you can surmise, no one has had this kind of money available in an experiment.

However, some researchers have been able to stretch the dollar a little bit in some other countries, or have been able to make $100 available (instead of $10) to a limited number of subjects. In one instance, two out of six people rejected offers of $30 out of $100 available. Altogether, while it makes sense that Player 2's are more likely to accept low-percentage offers with high stakes, the stakes probably have to be pretty high. In any event, we don't have any experimental evidence to know for sure.

Finally, does behavior differ according to demographic variables like gender, age, and so on?

> **NO BLUFFING**
>
> Since 2000 there have been a number of Ultimatum game experiments conducted mostly by anthropologists working with primitive tribal peoples all over the world. The Machiguenga in Peru come quite close to Nash equilibrium play, with low offers, virtually all of which are accepted. In a couple of tribes, including the Ache headhunters of Paraguay, average offers top 50 percent, reminiscent of the potlatch behavior I brought up in Chapter 8. But most of the tribal cultures offer in the 40 to 50 percent range on average, with 50 percent offers being the most common.

A number of studies focused on discovering gender differences. While some have hypothesized that women would play the Ultimatum game with less self-interest than men, the results don't exactly bear this out.

Does age have any effect on behavior in the Ultimatum game? There haven't been many studies, but the results so far are interesting. Young children offer less than older children. If this result is taken literally, it shows that younger children's actions are closer to subgame perfect Nash equilibria, and in this sense they become less competent game theorists as they get older. Taken more sensibly, the results show that either as their brains develop, children become more altruistic as they grow; or, as they get older, children learn societal norms, which punish excessive self-interest and reward generosity.

Other results show, as Camerer reports, that taller children offer less on average than other children. This is an interesting finding, especially when paired with data that show a correlation between one's relative height as a child and adult income.

To close out the Ultimatum game discussion, let me ask you whether you've ever noticed your own generosity in certain situations vary according to the physical attractiveness of the other person.

In one fascinating Ultimatum game study, Player 1's were shown photos of prospective Player 2's. While men did not tend to offer too much more to attractive women, women did offer considerably more, on the average, to good-looking men. In fact, women in this experiment offered more than 50 percent of the available sum, on average, to good-looking men. Overall, good-looking people were offered about 10 percent more (on average) than others, and men were offered about 15 percent more than women. Both of these findings correlate with studies in the workplace that show (a.), that there is a premium placed on beauty, and (b.), that men make more (when controlling for other variables) than women.

Coordination Experiments

Way back in Chapter 1 I introduced the Coordination game. Remember how Sony and Toshiba had to decide whether to jointly adopt Blu-ray or HD DVD? While each player had its own preferred outcome, the key sticking point of the game was that failing to coordinate at all led to a mutually poor outcome.

Starting in the 1990s, experimental game theorists have put this kind of coordination conundrum to the test. How do people behave in these situations, at least in the laboratory?

A series of coordination experiments was based on the following matrix game, with a payoff structure very similar to the one we used in Chapters 1 and 4.

Coordination Game Matrix

		Player 2 col. 1	col. 2
Player 1	row 1	(0, 0)	(200, 600)
	row 2	(600, 200)	(0, 0)

Payoffs in points

First, let's see what theory predicts. Without going through the math here, it turns out that the Nash equilibrium strategies are (¼, ¾) for both players. In other words, under this strategy pair, Player 1 will randomly pick row 1 ¼ of the time and row 2 ¾ of the time. Player 2 will adopt the same mix across the columns. As a result, the most frequent outcome is the (0, 0) at row 2 and column 2. In fact, it turns out that players employing the Nash equilibrium strategies will fail to coordinate ⅝, or 62.5 percent, of the time.

So much of what we've seen for the last few chapters has pointed out how players deviate from rational behavior. In this case, however, the subjects failed to coordinate 59 percent of the time, and were pretty much evenly divided between the other two outcomes. This is a remarkable match between theory and practice. Another researcher replicated the same experiment a few years later with very similar results.

You might be skeptical of this fit between the mixed-strategy Nash theory and the actual behavior. Maybe this was some sort of coincidence, perhaps having to do with the payoffs being symmetric. Several other versions of this game were run. Some versions gave one of the players an alternative, "outside" option with a guaranteed payoff. Other versions allowed the players to communicate: for example, Player 1 could say to Player 2, "I am going to play row 2." The results generally indicate that in these Coordination games, the players strategized almost according to the book.

> **BET ON IT**
>
> One variation in the Coordination game experiments that did not work out so well was, ironically, with two-way communication. With one-way communication, Player 1 would declare his preferred strategy and 96 percent of the time the players indeed ended up at (600,200). But with two-way communication, what mostly happened is that each player stated their preferred intention and this did nothing to increase their coordination.

One more worthwhile coordination study involves Stag Hunt, which we had a look at in Chapter 5. In Stag Hunt each player had two strategies: hunt stag, which means they cooperate toward a handsome payoff; or hunt rabbit, which means they run off after a smaller payoff. The person who plays stag while the other snares a rabbit will fare poorly.

The following payoff matrix was the basis for a series of Stag Hunt experiments.

Stag Hunt Matrix

		Player 2 stag	Player 2 rabbit
Player 1	stag	(1000, 1000)	(0, 800)
Player 1	rabbit	(800, 0)	(800, 800)

The payoff relationships differ from the ones I used in Chapter 5 in a significant way. To my mind, the rabbit payoff of 800 is too high relative to the mutual stag payoff of 1,000; such a high rabbit payoff does not capture the group benefit, and the resulting dilemma, of the original game. Playing rabbit in this instance guarantees a payoff of 800. While the mutual stag payoff of 1,000 is obviously higher, there is a lot of risk involved (the other player might play rabbit and then you're out of luck).

Sure enough, in the actual experiment, 97 percent of the outcomes were (rabbit, rabbit) with the payoff of (800, 800) and the other 3 percent of the time the players did not coordinate. In other words, almost all of the players realized that the guarantee from playing rabbit was too good to pass up. Another way to say it is that almost everyone played a maximin, or best of the worst, strategy here.

A later study explored whether the relative difference in the rabbit versus stag payoffs made a difference in how the subjects played. Sure enough, when the risk involved in playing stag was reduced past a certain tipping point, the players realized this and stopped playing rabbit.

It is too soon to draw conclusions about whether the findings from behavioral game theory confirm or disprove conjectures about human rationality. With Stag Hunt and the Coordination game, subjects played with a high degree of rationality. Rational play in the Dictator game calls for extreme self-interest, and the results we've seen do not diverge from that sort of selfishness all that much.

It's true that in the Ultimatum and Centipede games, people are much more altruistic than pure theory would suggest. My take on this is that it gives *homo sapiens* something to feel optimistic about. Unlike prospect theory, it is hard in the experimental games to get a consistent feel for how people will play. In the next chapter I round out our studies of behavioral games, and then you can draw your own conclusions.

The Least You Need to Know

- In Dictator games, people keep at least 80 percent of the money on average.
- In Centipede games and other back-and-forth exchanges, people show a fair amount of trust, and much of that trust is generally returned.
- In Ultimatum games, most players act strategically but also display a good deal of altruism.
- Most Ultimatum offers are in the 40 percent neighborhood, while offers in the 20 percent neighborhood are rejected about half the time.
- People deviate from the prescribed Nash equilibrium strategies in most of these behavioral games, but play very close to Nash predictions in Coordination games.

More Quirky Behavior

Chapter 21

In This Chapter

- Relying on dominance
- Putting mixed strategies to the test
- Thinking about what others think
- Testing the limits of rationality

In this chapter I'm serving up the second half of our introduction to behavioral game theory. There are still some unanswered questions about how people think, and more importantly, how they act in game situations. For example, standard theory says that dominated strategies are eliminated from consideration. Do people really behave accordingly?

Another foundation of game theory is the need to mix strategies in many situations. Does this occur in practice? Do people really take into account what others are thinking? Are there further cognitive discoveries that are relevant for us? Finding out the answers to these and other questions gives us a more complete picture of the relationship between theory and practice.

Dominant Strategies in Practice

For as long as I can remember, I've always looked both ways when crossing the street. Even when the street is obviously one-way, I still check to see if cars are coming from the wrong direction. Something tells me that maybe, just maybe, somebody's made a mistake (or is in a police chase) and they're barreling down the road unexpectedly.

It's the same with dominated strategies in games. Suppose you know the opponent has a dominated strategy. Can you count on them to discard it from consideration, or is there some chance they're going to play it?

Here's a pretty simple nonzero sum game from the experimental literature.

Picking a dominant strategy

		Player 2 Col. 1	Col. 2
Player 1	Row 1	(9.75, 3)	(9.75, 3)
	Row 2	(3, 4.75)	(10, 5)

Payoffs are in dollars

Although the game appears in strategic form, the players move sequentially in this perfect information game. Player 1 moves first. If he chooses row 1, he is guaranteed a payoff of 9.75. If he plays row 2, then it is Player 2's turn. Notice that column 2 dominates column 1, and in particular, once Player 2 finds herself restricted to the row 2 payoffs, then the payoff of 5 in column 2 is strictly better than the payoff of 4.75 in column 1. The strategy pair (row 2, column 2) with payoff (10, 5) is the subgame perfect equilibrium.

Suppose you are Player 1. Do you trust Player 2 to pick her dominant strategy? The thing is, if for some reason she plays column 1, then you are left with a payoff of only 3 instead of the 9.75 guarantee from row 1.

When the results of the experiment were in, 66 percent of the Player 1's played it safe, sticking with row 1. In other words, they didn't trust Player 2 to play her dominant strategy. But when Player 1's did play row 2, 5 out of 6 of the Player 2's did play column 2 as they "should" have.

A couple of things might strike you in looking at the payoffs. One is that if the risk for Player 1 was reduced, more 1's would play row 2. Another feature of the game is that Player 2 might play column 1 to spite Player 1; the reason for this is that the payoff of (10, 5) is so much better for Player 1. At (3, 4.75) 2's payoff is slightly lower but she might derive some utility from beating Player 1.

These observations did not escape the experimenters. They ran a number of different versions of the game, in which they adjusted the payoffs to reduce the risk. In one case they even tried increasing the resentment Player 2 would feel. When the risk to

Player 1 was reduced, 80 percent of the row players chose row 2, trusting Player 2 to play her dominant strategy.

> **BET ON IT**
>
> Even more noteworthy is the fact that in every variation of the game (other than the initial version) all Player 2's played their dominant strategy. Sometimes players trusted others to use their dominant strategy and sometimes they didn't; but almost always, when given the chance, Player 2's did in fact apply the principle of dominance.

Other experiments have tested whether people will use dominant strategies, and altogether the message seems to be that almost all players do recognize and act on dominance, but players are also not very confident that others will behave similarly.

Mixed Strategies in Practice

Imagine that you and 32 of your friends and family are congregating in a fenced-off field, and two chocolate-tossing robots appear on the other side of the fence and start shooting Hershey's Kisses (or maybe some Neuhaus chocolates) in gentle arcs at you. Robot 1 shoots out a chocolate every 5 seconds and Robot 2 shoots out a chocolate every 10 seconds. Where would you position yourself to maximize your chocolate haul, given that everyone else is out to do the same thing? Do you imagine yourself shuttling back and forth between the robots, and if so, how?

Frankly, I'm imagining a chaotic scene, with the 33 of you jostling each other and impulsively dashing around trying to improve your chances. If too many gather in front of Robot 1, then some will rush over to Robot 2, and so on.

We're supposed to be the smartest species on the planet, but sometimes you have to wonder. Here's something else to get you thinking: in 1979 a Cambridge University researcher did exactly this experiment—not with people and chocolates, but with ducks and uniformly sized balls of bread. The duck pond provided a captive audience of 33 ducks.

One person threw a bread ball every 5 seconds, another threw one every 10 seconds. Remarkably, in less than one minute, the ducks assembled themselves in front of the researchers precisely according to the Nash equilibrium solution of the game: $2/3$ of the ducks gathered in front of the faster bread-thrower, and the remaining $1/3$ of the ducks gathered in front of the slower one.

On top of that, when the researchers varied the size of the bread balls or varied the rate at which they tossed them, the ducks very quickly adapted to the new conditions and assembled themselves according to the Nash equilibrium mix. Okay, you reply, maybe we're not as smart as ducks; but already you're wondering whether people, in general, do play the mixed strategies that game theory prescribes.

Dozens of studies have tried to answer this question, and there are no clear-cut results. Colin Camerer shows that when all the studies are taken together, the averages of players' responses in the games are somewhat, but not strongly, correlated with the correct mixed strategy equilibria.

Some interesting studies have been done of tennis players' serves and soccer players' penalty kicks. If you remember, in Chapter 1 I used the penalty kick scenario as an example of a game in extensive form. A pair of studies in 2002 and 2003 showed that the payoff, measured as the probability of a goal being scored, is almost exactly the same when kickers kick right or left, or when the goalkeepers dive right or left.

The penalty kick game is zero sum—and remember, the key to an equilibrium is that the payoffs are equalized under either pure strategy. Since the probabilities of scoring a goal come out just about the same for kicking (or diving) left or right, this means that soccer players are playing their equilibrium strategies.

> **NO BLUFFING**
>
> You might wonder whether some kickers kick the ball straight, since that strategy would score a goal whether the goalkeeper moved left or right. In fact, this third strategy does make things more interesting. A 2007 study shows that goalkeepers would maximize their chances of blocking the kick if they did not dive left or right—and they are aware of this—but no one actually stands still because it then appears as if they didn't try hard enough!

Now we will switch gears a little bit and discover whether players anticipate what others are thinking to any deeper level.

Beauty Contests and Other Minds

Back in Chapter 15 I introduced the phrase "beauty contest" in the sense of companies sending in proposals for a contract and administrators picking the best-looking one. But now we will use the other sense of the phrase in game theory: when people try to anticipate what choices or rankings others make from a set of alternatives.

The term *beauty contest* itself comes from John Maynard Keynes, who used the following reasoning as a way to show that certain stock prices might go up or down as a result of the stock's popularity and not necessarily its economic fundamentals. Imagine an online contest where you look at 100 photos and are asked to pick the prettiest 5. The person who picks the five most-frequently selected photos is the winner.

A naïve way to play this game is to pick the five that you find prettiest. But then you'd realize that the ones you think are prettiest aren't necessarily the ones others might pick. Now you are one step ahead: you pick the five that you think everyone else would pick. But then you realize that everyone else will think this way, too; with this additional insight you're two steps ahead of the most basic play. You now aim to deduce which ones others believe that others would pick!

> **NO BLUFFING**
>
> Keynes believed that not only do many people go through these three "degrees" of anticipation, but that some people go well beyond: four, five, even six!

Just to make sure we have this straight, let's think about picking a stock. Here are the increasing levels of reasoning about what others are thinking:

- At the basic level, you simply pick a stock you like.

- At the next step, you pick a stock you think everyone else will pick.

- At the next step after that, you pick what you think everybody thinks everyone else will pick.

Exploring these steps of reasoning is certainly something that is relevant, even central, to game theory. So let's consider an experiment. Instead of looking at photos, suppose n people each privately pick a number from 0 to 100. The winner of the contest is the person who guesses the number that is $2/3$ of the average number picked.

- **Step 0:** The basic, gut response. At this level, you are not considering what others would do. Suppose you pick 50, just because it's an average number.

   Let's analyze this guess. The only way you could be the winner is if the average of the other responses was 75. Seeing this, I hope it makes sense that not only is 50 an unlikely winner, but that numbers above 50 are even less likely winners.

- **Step 1 (one step beyond):** What if everyone else picked 50? Then your best guess is to pick ⅔ of 50, which is 33.
- **Step 2 (two steps beyond):** If everyone else is pretty bright—maybe not as bright as you, but still, let's give people some credit—then they'd realize 50 is a poor guess. They will pick numbers a lot smaller than 50 (maybe in the 33 range), and then you'll have to pick a number that's ⅔ of that average.

I don't want to fill up the rest of the book going on and on to ever-higher levels of reasoning. I hope that you get the idea that as the numbers drop, then ⅔ of the average will drop too, and after repeatedly carrying out this reasoning, we would pretty soon get down to zero. In other words, if we carry on this reasoning infinitely many times (which is what ultra-rational players would do), the Nash equilibrium value is zero. But how do normal people play this game?

> **BET ON IT**
>
> No one picks zero in these games unless they're trying to impress the experimenters with how smart they are, or if they're at a game theory convention.

The first such experiment was performed in Germany. The average number picked was about 35. More telling, one big cluster of responses was in the neighborhood of 33 (one step reasoning) and another cluster was around 22 (two steps). By the way, in the literature on these levels of reasoning, the process of going from one level to the next and strategizing what to do based on what others would do at the previous level, is called *iterated dominance*.

Quite a few other beauty contest games have been run (some even appearing in newspapers) and the general sense is that most people employ one or two steps of reasoning beyond a basic response. It is dangerous to extrapolate these results to all manner of other games. However, one reasonable takeaway is that you can't expect many people in a game situation to be utterly naïve, and Keynes notwithstanding, don't expect many people to perform more than two steps of reasoning past a basic answer.

Prisoner's Dilemma Experiments

We've seen a lot of games in this book, and have come to realize that the Prisoner's Dilemma has remained the most fascinating and frustrating game of all. We know very well that the PD has a single, dominant strategy Nash equilibrium (when both players defect), which unfortunately leads to a poor outcome. How do people actually play the Prisoner's Dilemma? Do they hew to the game-theoretic prescription or do they cooperate? Does it make a difference when the game is played over and over?

Early Findings

The very first PD experiment was run in 1950, right after Merrill Flood and Melvin Dresher invented the game. That afternoon Flood and Dresher recruited two friends, economist Armen Alchian and mathematician John D. Williams, to play the game against each other 100 times in a row.

Both players wanted to elicit cooperation in the other party, and indeed they would go through stretches of mutual cooperation, but eventually one of them would break the string by defecting. The other one would punish that defection with one of his own, there would be some chaos for a while, and then another nice cooperative streak.

Altogether, Alchian cooperated 68 of the 100 times and Williams, 78. Not exactly Nash equilibrium play (which is to defect every single time). However, when they got down to the last play (and they knew it was the last play), guess what: they both defected.

NO BLUFFING

The best part of the very first PD experiment is the written transcript of the players' annotations as they proceeded. Alchian, for the most part, was dismayed by Williams's play. Besides calling Williams's play "perverse" at one point, Alchian's most telling remark was his recognition that Williams wanted more cooperation than Alchian was willing to give.

Williams, for his part, kept trying to train Alchian to cooperate by immediately punishing him when he defected. Williams's transcript showed a lot of exasperation with Alchian's tactics, calling him a "dope," a "knave," and a "shady character," as well as some deleted expletives.

Nash himself, when he heard about this experiment, wasn't impressed. From his point of view, once the players knew they were playing the PD repeatedly, the experiment was not about how they played a single PD many different times. Because each play was tied to the ones before, as well as the ones yet to come, for Nash the experiment involved one big game with 100 moves, rather than 100 separate games.

Since 1950 there have been tons of experiments using the Prisoner's Dilemma. One interesting finding published in 1970 and based on experimental data is that people can more or less be sorted into two camps: the cooperative and the competitive. This may not be terribly surprising in itself. But the interesting follow-on is that the cooperative types seem to recognize the diversity of their potential competitors' approaches to the game; competitive types, however, seem to believe that all their opponents are competitive as well.

While the cooperative/competitive dichotomy is a noteworthy discovery and it certainly provides food for thought, I don't want to paint with too broad a brush here. The crux of the PD is certainly the cooperative-competitive struggle. Let's see it put to the test.

Later Discoveries

Back in Chapter 5 we looked at strong and weak PDs, the idea being that if the fear of the other's defection as well as the greed in one's own temptation payoff could both be tempered, then people would be better disposed toward playing cooperatively.

In Chapter 8, I brought up Robert Frank's idea that the emotional side to our decision-making, while counterproductive in the short-term, might hold some long-term advantages by building up a reputation for being irrational. Associated with this emotional side is the question of being able to hide it; paradoxically, in situations requiring trust it can be to our advantage to show our true colors, so to speak.

Let's combine both these themes with respect to playing the Prisoner's Dilemma as a one-and-done game. Suppose we accept the results of the earlier research that show people to be either cooperative or competitive. If you are a competitive type, you expect others to defect and you will as well. But cooperators have a propensity to cooperate. How should they play?

While cooperators want to play cooperatively, if they know they're up against a competitive type their best strategy is to defect—and they will. But if they know the opponent is also cooperative, they will cooperate. In a 1993 study, the idea was to let the players interact with one another before they play the game. If people are any good at gauging others' personalities, they would be able to glean some valuable signals prior to the contest and then play accordingly.

If you were a cooperative type and were up against someone whom you couldn't get a read on, your dominant strategy (and maximin strategy, by the way) is to defect. But if the opponent seemed a sympathetic type, then you would be inclined to cooperate.

Note that defectors might try to mimic cooperators, but to do so would require some solid acting skill.

The game matrix, with payoffs in real money, is below.

One-shot experimental PD matrix

		Other player C	D
You	C	(2, 2)	(0, 3)
	D	(3, 0)	(1, 1)

Payoffs in dollars

The 99 subjects were split up into groups of 3. Each group spent a half hour alone prior to the PD play. They were allowed to talk freely in general and, in particular, could discuss strategies in the upcoming game. The strategies they played were kept not only confidential but also anonymous. Each player played a total of two games (one game against each of the two others in their group).

> **BET ON IT**
>
> When the subjects were to be paid what they earned in the games, the experimenters added random amounts to that sum so no subject would be able to figure out how the others in their group had played.

In addition to indicating on a form what their actions were in each of the two games in their group, subjects also indicated how they thought each of the opponents would play, based on their previous interaction. Along with the prediction, the subjects also wrote down a number from 50 to 100 indicating how confident they were in their prediction (50 meant no more confident than a coin flip, while 100 was totally confident).

Out of a total of 198 plays, there were 146 cooperations (C's) and 52 defections (D's). Overall, the subjects predicted cooperation 81 percent of the time, while the actual proportion of C's was 74 percent. The subjects predicted 130 of the 146 C's and 31 out of the 52 D's correctly for a 76.3 percent accuracy score. This score was much better than for a random guess from this population.

To conclude, even though subjects somewhat overestimated the proportion of true cooperators, in that half-hour period before play most subjects were able to figure out what kind of person they were up against and then play accordingly. Moreover, even though their anonymity was protected, only 26 percent of the subjects chose to defect. Both of these results are encouraging from the point of view of developing cooperation not only in the PD proper but in other mixed-motive endeavors.

Our Limited Abilities

Way before the revelations of prospect theory, Nobel Laureate Herbert Simon famously argued that we are far from the infinitely skilled and rational decision makers that economic theory and game theory demand us to be. His phrase, "bounded rationality," represents the imperfect tools with which we navigate the world of decision-making as well as the limits imposed by our environment. The following three points should make this clear.

- We are limited in the information that is available when we need to make decisions. This is nothing new to us; we've seen that game theory is rich in models of incomplete information. Yet we have to concede that a dearth of information certainly makes our decisions, well, less than informed.

- We usually operate with limited time with which to make decisions. This is clearly true and yet, you might protest that this point doesn't really speak to our own personal abilities.

- We have limited cognitive abilities, which I expand on below.

Not only do we have these three factors working against us, but they combine together and force us to come up with some shortcuts and *heuristics* to help ourselves make decisions.

> **DEFINITION**
>
> A **heuristic** is a rule of thumb, or an intuitive-based procedure, that people employ to solve problems. Heuristic procedures are usually simple to define and carry out; they do not involve elaborate formulas or steps of reasoning.

None of this is good news for those game theorists and economists who believe that people are rational decision makers. Over a period of decades, psychologists have conducted hundreds (if not thousands) of experiments that have studied our cognitive abilities and they've categorized our many shortcomings in the reasoning department. To name a few of our many human foibles that are relevant for game theory:

- We sometimes violate transitivity in our reasoning.
- We often fall victim to logical fallacies.
- Many of us don't understand statistical independence.
- We are poor at updating probabilities in the face of new evidence.

We've already talked about transitivity. But how many of us fall prey to the following type of blunder regarding independence: the weather forecaster says there's a 50 percent chance of rain on Saturday and a 50 percent chance of rain on Sunday, and we conclude that it will definitely rain this weekend.

POISONED PAWN

Don't confuse the deficiencies I'm mentioning here with the sorts of quirky preferences turned up by Prospect Theory (see Chapter 18). These are two different animals. Right now we're talking about limited cognitive capabilities, not preferences over risky items and lotteries.

How about one of many logical fallacies: you are in line at the movie theater and a 40-year-old guy who is about 6'7" is in front of you. Which is more likely: he is a stockbroker; or he is a stockbroker and a former basketball player? (Many people think the latter is more likely, but logically, it can't be.)

While decades of psychologists have had a field day shooting us down, I want to close out this chapter with a different take on our all-too-human tendencies. Psychologist Gerd Gigerenzer and others have been able to demonstrate that many of the shortcuts and heuristics that unconsciously guide our decisions are actually effective tools that have been adaptively shaped over tens of thousands of years (if not more) of evolution. In other words, many of the "gut feelings" that drive our decisions are in our decision-making toolbox precisely because they have proven to be effective.

One example of Gigerenzer's insight into the effectiveness of our heuristic approach involves cases similar to the stockbroker/basketball player above. For most psychologists, the example illustrates the so-called conjunction fallacy: there is less chance that two things can be true together (like being a stockbroker and a basketball player) than for one of them to be true. But many people get it wrong. Gigerenzer, however, looks more deeply into the origin of our intuition.

If you ask people what the word *likely* means, most will address the meaning as relevant to the context. Gigerenzer showed that people will define the word with such terms as plausible, reasonable, or typical; but they do not attach a mathematical sense to terms such as likely or probable.

Thus, the origin of our "logical" mistake is this: given that the salient information was the person's being 6'7", we interpret the question in the sense of what is reasonable as opposed to what is mathematically more accurate. So we latch on to the basketball player response and thus get the question wrong, logically speaking. But when terms like *probable* and *likely* are replaced by numerical, counting expressions like "how many," the fallacy disappears.

In other words, suppose we reconstructed the example as follows: out of every 100 men who are 6'7" tall, how many of them are (a.) stockbrokers, or (b.) stockbrokers and former basketball players.

In this sort of version, since people are counting and not making contextual associations, the vast majority in Gigerenzer's experiment (which used a different example) got it right.

There are other instances of such gut feelings and unconscious intelligence, but the main point is that our minds are adapted to looking at the context of problems. The associations we make and the conclusions we draw go beyond sheer logic, and our heuristic processes, evolutionarily speaking, have been very successful for us as a species.

The Least You Need to Know

- Players don't always trust others to use a dominant strategy, but dominance is almost always obeyed.
- Ordinary people mix strategies, but not optimally, while experts like soccer and tennis pros play practically according to the book.
- Players do anticipate other's moves, but not to advanced levels of sophistication.
- Prisoner's Dilemma experiments show a good deal of cooperation, especially when players are able to get a read on their opponents before playing.
- We exhibit more bounded rationality than full rationality, but some of our apparent limitations in fact provide for effective problem-solving.

Repeated Games and Tendencies

Chapter 22

In This Chapter

- The role of repetition in strategy
- Some irrationality can be rational
- Play nicely and carry a big stick
- Nice guys can finish first

This chapter is devoted to games that are played more than one time. It makes sense to study repeated games—aside from their theoretical interest—since we repeat our interactions so frequently in the real world. The first question that comes to mind is whether players should strategize differently when the game is repeated, as opposed to played only once. The second question is, does actual play in repeated games follow what theory dictates?

One of my objectives in this book is to show that game theory, like any other science, has something positive to offer humanity. One of the important messages up to this point has been how people have a tendency to play more cooperatively than rational theory would prescribe. In this chapter I explore this theme more fully, pulling together theory, practice, and a new tool: computer simulations. Happily, it turns out that often, cooperation can be the best strategy in the long run.

Why Study Repeated Games

Consider one of the many games in this book, like the Prisoner's Dilemma, Stag Hunt, or the Coordination game. What you've learned thus far is how to best play these games when you confront them. Now suppose you have to play one of these games twice in a row. So what? At first glance, it would seem that the best strategy when playing the game twice is simply to repeat the same strategy.

Stepping back for a minute, though, isn't it possible that you would want to employ a certain strategy in the first game so as to influence how your opponent plays the second time around? And if the game is repeated many times, not just twice, it is easy to imagine developing patterns of play—threats and rewards—that are designed to shape future interaction. We had a taste of this in Chapter 21 when we looked at the very first repeated Prisoner's Dilemma experiment.

The other feature of repeated games is that they take place over periods of time. In business and political situations, for instance, a good deal of time may elapse from one interaction to the next. Therefore, another function of repeated games is to model the relative impact of present versus future benefits and costs. Depending on how much future interaction is worth to you, one central issue is whether it's worth it to sacrifice gains in the present for rewards in the future.

The Chain Store Paradox

Nobel Laureate Reinhard Selten, who came up with the idea of subgame perfect equilibrium, proposed a game called the Chain Store Paradox, which is essentially our Upstart versus Goliath matchup (see Chapter 4).

> **BET ON IT**
>
> Let's recap what you know about Upstart versus Goliath. It's a game between a powerful chain store (Goliath) and a startup retailer (Upstart), which seeks to enter Goliath's market. Upstart moves first and can either enter the market or stay out. If Upstart enters, Goliath can either play hardball (fighting Upstart tooth and nail with an advertising and price war) or can accommodate Upstart by not playing competitively.

In Chapter 4 the game used the following payoff matrix, which is the same as Selten's. Even though the game is technically in extensive form, I've used the strategic form (matrix) representation.

Suppose, prior to play, Goliath informed Upstart that it would play hardball if Upstart entered the market. This threat is designed to scare Upstart off. We realized that the threat is not credible because if Upstart enters, Goliath is better off playing accommodate. Such is the logic of subgame perfect equilibrium.

Upstart vs. Goliath, revisited

		Goliath's choices	
		Hardball	Accommodate
Upstart's choices	Enter	(0, 0)	(2, 2)
	Stay out	(1, 5)	(1, 5)

Strategic form representation

So what's the paradox? Suppose the game is played more than once—for example, if a series of Upstarts challenges the ubiquitous Goliath in 20 different geographically dispersed markets, one at a time. If Goliath treats the 20 different challenges independently—where no one result affects the outcome in any other market—then the logical strategy is to accommodate each Upstart as they enter. Notice that since Goliath's payoff is 2 when Upstart enters and the Goliath accommodates, then over 20 games the total payoff to Goliath is 20 * 2 = 40.

Before getting to the paradox, let's think of the series of 20 games as one big *supergame*. Looking at this one game with 20 plays, a series of accommodations by Goliath is validated by the sort of backward induction argument we've been getting used to. In play number 20, the strategy pair (enter, accommodate) is the unique subgame perfect equilibrium.

DEFINITION

A **supergame** is a game with repeated, identical plays, where each play is itself a well-defined game. The game that is repeated is typically one of the standard two-person games (like PD, etc.).

Now what about play 19? You know that Goliath will accommodate in play 20; in other words, whatever happens in play 19 makes no difference in play 20. In play 19 the equilibrium is once again (enter, accommodate). Go back to play 18, and the same argument repeats itself, then the same for play 17, and so on, all the way back to play 1. Thus the equilibrium in the supergame is (enter, accommodate) every single play.

As you can see, the rational strategy for Goliath in the repeated game is simply to accommodate the Upstarts one by one. But Goliath still dreams of scaring off potential Upstarts. Aside from verbal threats, which lack teeth, is there a way to dissuade the Upstarts from entering into the markets?

In Chapter 6, using a model of incomplete information, you saw that if an Upstart believed that Goliath was a "strong" player—one who derived utility from playing hardball—then the Upstart would be better off staying out. In this vein, it might be possible for Goliath to play hardball in play 1, and hope that walking the walk, so to speak, would deter future entrants. Can such a deterrence strategy work?

Selten came up with some specifics regarding deterrence. He assumed that in very late plays of the game—plays 18, 19, and 20—that the Upstarts will realize there are not enough future competitors to be scared off. In other words, in plays 18, 19, and 20, the Upstarts will boldly play enter and Goliath will indeed accommodate them.

But Selten argues that when Goliath comes out fighting in the first few plays, then on the whole, it is unlikely that more than five or so of the Upstarts before round 18 will enter when they see their predecessors getting hammered. If Goliath accommodates the three entrants in plays 18 through 20, but plays hardball against several entering Upstarts in the early going, and has the remaining 12 or so Upstarts in the middle rounds chicken out, then Goliath will have collected a total payoff of, for example,

$$5 * (0) + 12 * 5 + 3 * (2) = 66$$

which is far better than the 40 from a series of accommodations. The several zero payoffs for Goliath in the early going are more than made up for by the dozen payoffs of 5 points each in the middle rounds.

To sum up the paradox: using rational decision-making in the repeated game, Goliath will accommodate a series of 20 entrants and obtain a total payoff of 40. But scaring off most of Upstarts by playing hardball early on will rack up, say, more than 60 points. Which theory do you subscribe to? Selten himself, while indicating that only the backward induction theory is sound, is nevertheless squarely in the deterrence camp.

As you've seen before, mathematical rationality is one thing; but the deterrence strategy here would seem to be effective against human Upstarts who would draw the following two conclusions:

- If Goliath has played hardball a few times in a row, they will probably play hardball again (unless it's the last couple of plays).
- If the Upstarts were in Goliath's shoes, they, too, would play hardball and similarly expect to scare off most of the competition.

Repeated PDs

Even Selten in his Chain Store article couldn't resist the urge to start talking about the Prisoner's Dilemma. So let's return to the PD once again. When the Prisoner's Dilemma is played just once, clearly the rational strategy is to defect. But what about when it is repeated many times?

Pure, unadulterated game theory states that a Prisoner's Dilemma repeated any finite number of plays will have the players mutually defecting every single time. This is found using the backward induction argument. In the very last PD play, the Nash equilibrium strategy is mutual defection. In the second-to-last game, same thing, and so on, all the way back to the very first encounter.

How do real people play repeated PD's? In experiments run over the years, lots of players frequently cooperate. Of course, this has been somewhat embarrassing to many theorists, and finally in 1982 a group of eminent researchers called the "gang of four" found a theoretical way to justify cooperation as a rational strategy in the repeated PD.

The idea is that each player believes there is some probability p that the other player is irrational, meaning that they would cooperate in any one game. Maybe you think the other player will employ the "tit-for-tat" strategy you saw in Chapter 17, where they start out by cooperating and from then on, repeat what you did in the previous game. Or maybe, similar to what happened in Chapter 5, the payoffs to the opponent are different from yours. Maybe they get a deep, satisfying feeling from mutual cooperation.

Suppose the game is repeated many times, and that your opponent might be willing to cooperate but won't tolerate a defection from you. Let's say that to punish your defection, they are prepared to defect for all the remaining games. You realize that the long-term payoff from repeated cooperation exceeds that from repeated defection.

In this scenario, it is then rational for you to cooperate, at least early on in the repeated interactions. The theoretical result is that in this situation, with a high enough probability of playing the repeated PD with an irrational (cooperative) type, there is an equilibrium in which both players will play a series of mutually cooperative games, although not right up until the last one.

Incidentally, nobody said these irrational types actually exist; it is enough to merely believe that they exist. Even a pair of players who are normally inclined to defect (because to do so is rational) might go along and cooperate for a while.

If two players were to discover they were both rational, they would end up defecting from that point on. But now imagine that some cooperative types really do exist: these people would be expected to extend their cooperative streak longer than those inclined to defect.

This elaborate theory is based on two assumptions:

- Players will cooperate if they think there are cooperators out there.
- Reputations are built and acted on over the course of the encounters.

In other words, presupposing that irrational, cooperative players lurk here and there opens the door to a theory that supports cooperation.

Some Experimental Corroboration

What I've just sketched for you was put to the test in an experiment carried out in the 1990s. In this experiment, subjects played repeated PD against either the same partners, or against strangers, for 20 rounds of play each. Every round consisted of 10 PD plays.

If players were interested in building a reputation, such behavior would only take place when playing the same partner; the whole idea of having games against strangers as well is that if the games are played anonymously, then the whole idea of reputation is moot.

> **NO BLUFFING**
>
> Players playing repeated PD against either the same partner or against a field of competitors need to observe past results in order to update their prior probabilities, or personal "guesstimates," of the likelihood of having someone cooperate on the next play. In game theory, these updated prior probabilities are called "homemade priors."

The overall conclusions of the experiments validate the gang of four's theory. Even when playing against strangers, subjects cooperated a significant proportion of the time (about 20 percent, and at around 10 percent even on the last play). But when playing against the steady partners, the subjects achieved significantly higher co-operation levels (just over 40 percent).

Another thing the experimenters did was to set up games where subjects were told there was a (random) 50 percent chance that the partner was a computer that played the "tit-for-tat" strategy mentioned previously. This program would cooperate and then mimic the opponent's previous move. If you play nice against tit for tat, it will play nice against you.

The idea of throwing in random, well-behaved computer programs is that the subjects would be expected to play more cooperatively because the computer program is more predictable, and therefore more trustworthy than a wily human opponent. Sure enough, the subjects in the 50-50 setting indeed played more cooperatively.

Finally, judging from the patterns of play, the experimenters were able to place players into three groups. One group was the cooperators. These people cooperated in general, though usually not all the way through. Many cooperators, however, displayed unusual amounts of altruism.

Then there were the defectors. Guess what—they pretty much just defected. But then, as theory predicted, there were the "mixers," whose cooperation rate on average was about 20 percent. These players cooperated to some extent to develop a reputation for being nice, but defected early on, thereby reverting to their true colors.

Altogether, the following conclusions can be drawn from this set of experiments:

- When subjects played strangers, their cooperation level was at about 20 percent on average.
- When people played repeatedly against the same opponent, their cooperation level rose significantly, to higher than 40 percent on average.
- When people had a 50 percent chance of playing a dependable computer program that would not defect on them first, they cooperated even more.
- Some subjects showed true altruism in their patterns of play.
- The subjects could be fairly neatly divided into cooperators, defectors, and a third category called mixers.

Yet once again, behavioral game theory has shown us that a significant proportion of people do not assume perfect rationality in others; many respond favorably to cooperative play from others; and some have an innate altruism that is not neutralized by the lure of points or money. I receive this news with guarded optimism. But how can we reconcile the following two conflicting directions for game theory?

- Pure unadulterated game theory does not allow for opponents who are less than coldly rational. Consequently this theory states, for example, that players will strictly play Nash equilibria.

- Other theoretical models do allow for the possibility of irrational opponents. Once we get past this slippery slope, these theoretical models indicate that irrational strategies can be successfully employed.

Translation: cooperative play in the PD can result in a higher average payoff than unwavering defection.

Behavioral game theory shows that people frequently play according to the second model as opposed to the purely rational first model. In other words, many people display a cooperative, if not always an altruistic, streak.

BET ON IT

You already have seen evidence in the Ultimatum game (see Chapter 20) that many people are truly altruistic at heart. The results in this repeated PD study support that conviction.

Of course, there's not enough space for us to dissect the entire literature on repeated Prisoner's Dilemma games in the laboratory. But this study not only fits a reasonable and highly regarded theory, but also provides some pretty typical overall results. Time and again researchers observe more actual cooperation in the PD than rational theory predicts.

By now you have seen a number of games in which payoffs obtained through rational play are markedly inferior to those from alternative strategizing. In the Centipede game in Chapter 20 as well as in Upstart versus Goliath, the backward induction argument, while mathematically sound, yields payoffs that are worse than those from "irrational" play.

I am putting "irrational" in quotes because, given the evidence available, there is no sense in continuing to insist that such alternative strategizing is second-rate. The results I've presented to you challenge the very foundation of rationality in the context of strategic interaction.

The Folk Theorem

In the context of a repeated game, theorists have known for a long time that there can exist different sequences of strategies, each of which results in a subgame perfect equilibrium. These strategies can provide a wide range of payoffs, and since some are better than others, it becomes difficult to evaluate the strategies as to their "rationality."

This development is important, and consequently game theorists formulated a more precise way to express it, known as the folk theorem. Although there are different statements of this result, all of them share the same idea. To understand it better, let's use the PD as the context and assume it is repeated an infinite number of times. Recall that the maximin payoff occurs with mutual defection.

Loosely speaking, the folk theorem says that in a repeated game, any series of plays that yields an average payoff at least as good as the maximin payoff can be maintained by the players and is therefore a subgame perfect equilibrium, provided that future interaction is important enough. Players maintain this equilibrium by threatening to punish a defector.

More succinctly, the folk theorem says that for every individually rational payoff to both players, if future payoffs are important enough, there is a subgame perfect Nash equilibrium that corresponds to that payoff.

What this means for us plain and simple is, in a repeated PD for example, mutual cooperation can end up as a Nash equilibrium provided that future payoffs do not decrease too much in value as perceived at the beginning of the game. The equilibrium is enforced by both players' ability to carry out the threat of defecting forever after if the opponent deviates from the equilibrium strategy.

NO BLUFFING

It's called the folk theorem because it was something of oral tradition, or folklore, among game theorists for years before anyone published it as a formal result. Various researchers published different versions of it in the 1970s and 1980s.

Let's look into this a little deeper. First, imagine that you and a dance partner have choreographed a dance. A series of steps is involved: you move left, right, right again, backward, forward, whatever. The way the dance works is that you both perform the correct cycle of steps and, when that cycle is finished, you repeat it.

In the repeated PD, the "dance" is a series of moves (strategy pairs) of the two players, each with their associated payoffs. These moves can involve both cooperating (*C*'s) and defecting (*D*'s). As long as this series provides an average payoff that is at least as good as mutual defection, it can be maintained and is therefore an equilibrium.

As I've said, this arbitrary sequence of plays can be maintained when each player is prepared to activate the following threat against the other. If one of the players gets greedy and plays one defection too many, the other player will employ a so-called trigger strategy: to defect forever after.

Now here's the key thing: suppose that the future isn't worth much to the players. For example, if the rate of inflation were extremely high and the payoffs didn't get adjusted for it, then future payments aren't worth very much. If the future isn't worth much, then the trigger strategy won't be an effective deterrent against defection.

But if the future is worth a lot to the players, the trigger strategy will work. A player who cheats by defecting one time too many will then get hammered for the rest of the game. This resulting series of opponent's defections will drop the instigator's average score below what they could have attained had they behaved themselves in the first place.

> **BET ON IT**
>
> The actual mathematical argument to formally prove the folk theorem is somewhat advanced. There are also different versions of the theorem—for starters, the story is slightly different depending on whether you have an infinitely repeated game or a finitely repeated game. In the latter case the argument has to be modified a bit. But what I've presented here illustrates the essential idea.

How Cooperation Can Evolve

The Prisoner's Dilemma catches everyone's eye because it not only seems to capture a lot of mixed-motive interactions in our social and business worlds, but also is something of a metaphor for life itself. Some say that nice guys finish last. Indeed, rational play in the PD dictates that nice guys will be steamrolled by the meanies.

But the gang of four opened the door, theoretically, for cooperation to emerge. And you've also seen that real people, at least in laboratory experiments, do indicate a willingness to cooperate that goes beyond any semblance of rational theory. Now let's see how other researchers have utilized a new tool—computer simulations—that further help us understand the nature of cooperative play and how it can emerge.

Axelrod's Experiment

In the early 1980s political scientist Robert Axelrod was bent on answering the simple but deep question: if two people interact repeatedly, when should they cooperate and when should they be selfish?

To answer that question, Axelrod organized a computer tournament. He invited 14 experts in game theory to each submit a computer program for playing the Prisoner's Dilemma repeatedly. Along with a random program thrown in the mix, these 14 programs played 200 rounds against each of the others. (Each played its own "twin" as well.)

Using a tournament allowed Axelrod to compare strategies as they competed against one another. This provides a much more interesting landscape than just compiling data from one-on-one laboratory experiments. Now there are winners and losers.

The payoff matrix that Axelrod used was the standard one that we have seen. I've included it here as a reminder. Cooperating is indicated by C and defection is denoted by D.

Axelrod's PD matrix

		Player 2	
		C	D
Player 1	C	(3, 3)	(0, 5)
	D	(5, 0)	(1, 1)

Payoffs in points
Points earned in each round were accumulated

Points earned in each round were accumulated.

Notice that in any one interaction of 200 plays, the possible scores found by adding up all the payoffs range from 0 to 1,000. Both extremes would be unlikely. More reasonable totals would be somewhere from 200 points obtained from a series of mutual defections to 600 points, found from a series of mutual cooperations.

Imagine programming a computer to play PD. You could aim for simplicity or complexity. At one extreme, you could program the computer to defect (or cooperate) every single time. At the other extreme, you could collect data on every interaction with every other program, look at the patterns of play, update the probabilities of the opponent playing C or D, and adjust your strategy accordingly.

The game theorists submitted a variety of strategies for this tournament. The winner was the simplest program: Anatol Rapoport's code played tit for tat (TFT). It started out by cooperating and from that point on, played *D* if its opponent played *D* last time and played *C* if the opponent cooperated last time.

This result is interesting and important enough. But Axelrod was able to uncover much more. TFT averaged 505 points over the 200 rounds, below the mutually cooperative result of 600. Since some programs defected a lot in general, the low-sounding score is not surprising. Interestingly, the totally random program was at the bottom of the results table, with an average of 276.

What about the other programs? First, let's develop a simple way to categorize the programs using the cooperator-defector dichotomy we've already seen. Axelrod defined some programs as *nice* while the rest were *not nice*.

> **DEFINITION**
>
> In Axelrod's PD experiment, **nice** programs were defined as those, like TFT, that started out by cooperating, while programs that had any chance of starting out by defecting were **not nice**.

Looking at the results, there was a big gap separating the competitors. The top eight programs amassed at least 472 points, while the bottom seven programs were all below 401 points. Perhaps the biggest news in the entire study is that every one of the programs in the top half was nice, and every single program in the bottom half was not nice!

It is instructive to see what some of the other programs did. One program ("Joss") basically played TFT but every once in a while threw in a random defection. This sounds clever, because the random defections garner 5 points if the other player trusted you to cooperate. But in the long run, those few one-shot victories will be offset by punishments from other players.

Another program employed a sophisticated Bayesian updating mechanism. The idea was to look at the previous patterns of play with each opponent and to try to figure out if the opponent was responsive to what it was doing. If the opponent was a cooperative type, then the program would cooperate; if the opponent was an unresponsive defector, then the program would defect.

One thing Axelrod noticed was that TFT had the potential to get dragged down into a retaliatory war that would continue to the bitter end. If TFT played Joss, for example, look what's going to happen. First, both play *C* for a while. Joss throws in that random defection; TFT retaliates with a defection next round and that will set off an alternating series of *C*'s and *D*'s.

Sooner or later, though, Joss is going to throw in another random defection instead of playing *C*. In that round both TFT and Joss will have played *D* and then they both defect for the remainder of the contest. Both Joss and TFT will end up with scores well below the average of 3 that they should have been able to maintain if Joss didn't have that nasty urge to cheat occasionally. The story line looks like this.

TFT:	*C*	*C*	*C*	*C*	*D*	*C* ... *D D D* ...
Joss:	*C*	*C*	*C*	*D*	*C*	*D* ... *D D D* ...

BET ON IT

Another thing to notice is that TFT never beats any other competitor in terms of overall score. The only way for TFT to accumulate more points than an opponent is to have more *D*'s, but that's impossible. This makes it even more remarkable that TFT could end up the tournament winner.

Because TFT is prone to this sort of Hatfield-and-McCoy pattern of revenge, Axelrod realized that a more forgiving strategy might avoid getting into such a rut. It turns out that in this particular experiment, had someone entered "tit for two tats," it would have won.

Next, Axelrod set up a second tournament. This one featured 63 entrants. To avoid the backward induction problems inherent when the number of plays is finite, Axelrod rigged the sequence of PDs so that on any one play there was a tiny random chance (about 0.3 percent) the interaction would end. In other words, no program, just before playing another round, would be able to tell whether, or how long, the repeated game would go on.

In this expanded tournament the average length of interactions between pairs was 151 rounds. Guess what: TFT won again! Once again, nice guys finished first. Of the top 15 rules in the results table, all but one were nice, and of the bottom 15 rules, all but one were not nice.

I should interject that the results table reveals more than who won the tournament. Remember that the Prisoner's Dilemma payoff structure actually represents what businesspeople perceive when they enter into joint ventures or advertising wars.

Moreover, matrix games such as PD and Chicken even represent strategies and outcomes taking place in the wild. In other words, those seemingly innocent points appearing in the matrices are more than metaphors for profits and costs in the real world.

To model evolutionary pressures, Axelrod created a repeated tournament where more successful programs would have more offspring—representatives in the next generation—than less successful programs. In this evolutionary setting, by the fiftieth generation, the programs originally in the bottom third (mostly not nice ones) had virtually disappeared, while those in the top third were starting to take over this virtual world. The nice programs were pushing the mean ones out of the picture!

Finally, let's recall the point I made when discussing the folk theorem, about the importance of future interactions. If future interactions are negligible, then it doesn't pay to cooperate now. Axelrod was able to demonstrate this principle for strategies in his environment.

When I discussed the folk theorem, I measured the importance of future payoffs by how little they declined over time, for example, from the effects of inflation. Axelrod measured the importance of future interactions not by discounting the payoff amounts but by examining their likelihood.

Suppose that p represents the probability that you will play the next round against the opponent. (In Axelrod's second tournament, p was roughly 0.997.) What would the stream of future payoffs look like? Just to do one example, suppose two programs are constantly defecting against one another. The stream of expected payoffs would be

$$1 * p + 1 * p^2 + 1 * p^3 + \ldots$$

Even though this sequence goes on forever, the payoffs get smaller and the sum of the terms is actually finite. Suppose $p = \frac{1}{2}$. Then the sum of future payoffs is $\frac{1}{2} + \frac{1}{4} + \frac{1}{8} + \ldots$. This stream of expected payoffs adds up to exactly 1. If you could defect against a cooperator now and obtain 5 points, and future rounds were unlikely to occur, your best strategy is to defect.

Axelrod realized that TFT proved to be a robust strategy. But in his tournament, the next interaction happened with more than 99 percent probability. What if the future didn't loom so large? Could a mean program like all-D (always defect) invade a population of TFTs, in the sense of an evolutionary game that we studied in Chapter 17?

It turns out that if the future is important enough, all-*D*'s cannot successfully invade a population of TFTs. In fact, for the payoffs in the standard PD matrix, as long as the probability of a future interaction is at least ⅔, then TFT is "collectively stable"; that is, a population of TFTs can withstand an invasion of all-*D*s. In other words, provided the future is important enough, TFT is an evolutionarily stable strategy (ESS).

On the other hand—and this last result is a fitting way to end our survey of Axelrod's work—a small band of TFT's can successfully invade a population of all-*D*s (provided the future is important enough). Remember that when TFT plays all-*D*, TFT loses. But the all-*D*s do poorly among themselves, too, while the TFTs cooperate against other TFTs and this boosts their point totals.

POISONED PAWN

Although Axelrod's work probes deeply into the very question of how cooperation can emerge in a competitive environment, be careful about taking the results too literally. It might warm your heart to see "nice" programs triumph, but this does not mean that every single repeated game played in the real world will necessarily have a happy ending.

In terms of gathering points, if enough of the TFT's interactions are among like-minded players, the 3-point payoffs they receive from mutual cooperation will outweigh the 1-point payoffs from mutual defections among the all-*D* population. Nice guys can finish first!

Building Future Relationships

The results we've whizzed through continue to fill me, as I've said before, with guarded optimism. The theoretical work I presented to you earlier in this chapter shows that a little bit of cooperation and altruism in a society can go a long way toward fostering cooperative behavior, even in a thorny game like the Prisoner's Dilemma. Axelrod's tournaments, which have inspired a lot of "agent-based" computer simulations that are still being carried out today, validate the point.

In various places in this book we've seen how tricky conflicts can be handled. We've seen how difficult games like Chicken and Stag Hunt can be largely stripped of their fangs. Even the Prisoner's Dilemma doesn't strike us as an intractable struggle anymore.

Our study of repeated games provides an ideal way to wrap up our introduction to game theory. We've learned that building relationships can make a positive difference in the world. We know that our actions today can affect how we're treated tomorrow—as long as we're granted that tomorrow.

An old French proverb says that "friends of my friends are my friends." We can do more than hope that the world becomes a friendlier place; we can each play our own part. Sometimes it seems like the bad guys are winning, but at such times we have to remind ourselves that even a small band of cooperators can keep the bad guys at bay. As for developing friends, in the past several years social networking sites like Facebook have exploded all over the world.

It's too soon to tell whether social networking or other tools and forums will actually develop global cooperation in a meaningful way. But in my book, anything that might foster trust and break down suspicions is certainly worth a try.

Like other game theory authors who have preceded me, I'm getting a little sentimental as I'm saying goodbye. But it's my sincere hope, now that you've gone through this book, that you will be able to approach your world of social, personal, political, and business situations with that much more confidence and acumen.

The Least You Need to Know

- In the Chain Store Paradox, the only rational outcome is (enter, accommodate); but if Goliath plays hardball early on, future Upstarts will probably be scared off.
- The gang of four built a rational theory based on the premise that a small proportion of players are irrational.
- In experiments, many people cooperate in the repeated PD and their cooperation increases the more they think the opponent will cooperate.
- The folk theorem shows that cooperation can be an equilibrium in the repeated PD as long as the future is important and a threat strategy is in place.
- Axelrod's computer tournaments show how cooperation can thrive in a competitive environment.

Glossary

Appendix A

additivity   A property of functions like the characteristic function in a cooperative game. This function is additive for two disjoint, or separate, coalitions when the union of the coalitions obtains exactly the sum of their separate outcomes. Additivity implies there is no synergy between the two coalitions.

adverse selection   Situations in which information asymmetry leads to undesirable outcomes. The typical applications are in insurance and banking, where one party is informed of certain risks while the other is not.

anchor   A prior, numerical point of comparison.

asymmetric information   Occurs in games when one player knows the payoffs for both players but the other player only knows his or her own payoffs.

auctionlike mechanism   A procedure where ever-rising prices are proposed to players who are bidding to participate in a collective enterprise. The objective is to elicit honest revelations from the players.

average   The result of adding the numbers in a set and dividing the sum by the size of the set.

axiomatic   An approach to game theory that involves setting down some incontrovertible principles that any fair (or "reasonable") solution ought to satisfy and then finding a solution that works.

backward induction   A reasoning technique in which one starts with the final outcomes and reasons back to identify the best strategies to pick at the outset.

bargaining game   Game that involves two players and a subset of possible outcomes, represented as two-dimensional coordinate points, one of which must be selected as the solution.

Bayes' rule   A technique in which a prior probability is updated to a posterior probability through analyzing test results or other newly acquired information.

Bayesian games Games of incomplete information in which we model uncertainty with a move by nature.

Bayesian-Nash equilibrium A Nash equilibrium in a Bayesian game.

beauty contest In game theory, a process where decision makers award a contract after viewing a number of proposals and picking the "best-looking" one; also, a decision maker's choice of the "best" alternative is the one he or she believes others think is best.

best reply A strategy in which a player optimizes his or her play in the face of the opponent's strategy or strategy mix.

bounded rationality Concept meaning that we can't make perfectly rational decisions because the information we have at our disposal is incomplete, we are often short of the time necessary to make a proper decision, and most important of all, we have limited cognitive abilities.

brinkmanship The art of coercing your opponent by linking the strategies of both players to a risky move of nature.

budget balanced A set of payments made among a group of players together with an administrator in which the sum of the payments made by the players is at least as much as what is paid by the administrator. Budget balance means that there is no net deficit.

Centipede game A two-person game of perfect information in which players take turns either claiming a given proportion of a pot of money or passing a bigger pot of money to the other player.

certainty equivalent Given a particular utility function and a lottery, the sure payoff amount that is judged to be equivalent in utility to the lottery.

characteristic function In a cooperative game, it records every possible coalition that can form along with what that coalition can obtain for itself in the absence of the remaining players.

Chicken A two-player, nonzero sum game. Each player has two strategies: to be aggressive or passive. Being aggressive pays off against a passive opponent, but when both are aggressive the outcome is a disaster.

coalition A subset of players in a cooperative game.

common-value auction Auction in which the bidders do not even know what the item is worth to themselves, let alone what it is worth to others.

complete information   Games in which the players know both the strategies and the outcomes for all players.

constant sum game   A two-player game in which every possible pair of numerical outcomes adds up to a fixed number. Like zero sum games, in constant sum games what one player gains, the other loses.

cooperative game   Any game used to model situations in which players are better off when they join up with others. Cooperative games typically have more than two players, and the players benefit by forming subsets called coalitions.

coordination game   Two-player games characterized by acceptable (though not necessarily equal) payoffs when the two players' strategies match but unacceptable payoffs when their strategies do not match.

core   An efficient solution for a cooperative game that is individually rational and group rational.

decision criterion   A rule on which a decision is based. Two commonly used decision criteria are based on expected values or a maximin approach.

decision theory   When the opponent is nature, and not another player, a theory that encompasses different methodologies that select the best alternative based on comparisons using expected values or other criteria.

decision tree   A branching diagram that indicates different alternatives, random events and their probabilities, and numerical outcomes.

descriptive   Aspect of game theory that shows what players actually do in real-life game situations.

Dictator game   Game in which there are two players and a sum of money. One player decides how the money is to be divided. The other player performs no action in the game.

disagreement point   In a bargaining game, a fallback (maximin) outcome, usually (0, 0), where the players end up if they cannot come to an agreement.

discount rate   An annual measure of how the value of money changes over time, usually due to interest or inflation. For our purposes it is the same as an annual interest rate or rate of inflation.

dominant strategy   A strategy that dominates all other strategies for a player.

domination Occurs if, when comparing two strategies A and B, every outcome under A is better for a player as the corresponding outcomes under strategy B.

dummy players In cooperative games, players who generate no value when acting alone and do not add any value to any other coalition.

efficient In game theory, a type of solution that maximizes the total benefits to the players.

English auction Type of auction in which the price of the item keeps increasing and bidders drop out until there is only one left. That bidder wins the object at the last announced price.

envy-free solution A solution in which all players like their own allocations at least as much as any others.

equilibrium A pair of strategies such that neither player will improve his or her payoff by (unilaterally) deviating from their strategy.

equitable An assignment of items to players in which each player thinks that what he or she gets is worth the same to him or her as what the other player gets as valued by the other player.

evolutionarily stable strategy (ESS) A strategy which, if pursued by a population of individuals, will protect them from invasion by others using any alternative strategy.

expected value A weighted average in which each number is multiplied by its associated probability.

extensive form A diagrammatic representation of a game using branches to depict the players moving one after the other.

fair division Type of problem with two or more players and with a number of items, including indivisible ones, to be allocated or divided among them.

fitness In evolutionary biology (and evolutionary game theory), the relative ability of an individual to survive and pass on its genes.

folk theorem In repeated games, theorem stating that for every individually rational payoff to both players, if future payoffs are important enough, there is a Nash equilibrium that corresponds to that payoff.

free rider A player who shares in the overall benefit obtained in a game but does not contribute to the cost of providing the benefit.

frequency Given a set of numbers or outcomes, how many times certain outcomes appear in the set.

game A situation in which there are multiple decision makers. Each decision maker is called a player, and each player has a certain set of strategies available to him or her. When each player settles on a particular strategy, the result is called an *outcome*.

game theory A mathematical approach to analyzing conflict among two or more decision makers.

generalized second-price auction An auction that has at least one item, and each item's winner pays the amount of the next-highest bid for that item. In these auctions, honest bidding is not necessarily a dominant strategy.

group rationality A condition in cooperative games in which every possible coalition is able to obtain an outcome that is at least as good as can be obtained when they act in the absence of the other players.

handicap principle States that very costly signals will be reliable. For example, if a player is clearly able to shoulder a great burden or cost, then that player has proven their strength or fitness.

heuristic A rule of thumb, or an intuitive-based procedure, that people employ to solve problems. Heuristic procedures are usually simple to define and carry out; they do not involve elaborate formulas or steps of reasoning.

imperfect information Games in which players do not know what the others have done when it is their turn to move.

incentive compatible Games of incomplete information in which the Nash equilibrium is found when all players play truthfully, i.e., honestly reveal their valuations.

incomplete information Games in which the players might know their own outcomes but do not know the outcomes for the other players.

independence of irrelevant alternatives In a bargaining game, if R is a subset of S and if the solution for S is a point in R, then this solution must solve the bargaining game defined for the smaller set R.

individually rational Solution that ensures that each player obtains an outcome in the game that is at least as good as can be obtained by opting out of the game.

information set A grouping of nodes in an extensive form diagram. The circumstances at the different nodes might be very different, but the player who is to move cannot distinguish among them.

irrational Behavior that does not follow the strategy that optimizes one's utility, or preferences that violate von Neumann-Morgenstern utility.

lottery A set of alternatives along with a probability distribution over those alternatives.

mathematical model The use of variables, equations, probabilities, and so on, in order to represent a real-world situation.

maximin A decision rule in profit situations where one selects the alternative yielding the best of the worst-case outcomes.

mechanism design A way to combat problems caused by private information; the idea is to create a mechanism, i.e., engineer the rules of the game to get the players to behave in a certain way.

minimax A decision rule in cost situations where one selects the alternative yielding the best of the worst-case outcomes.

mixed strategy Occurs when a player selects a strategy from a set of choices according to a probability distribution on those choices.

mixed strategy equilibrium An equilibrium in which all players are using mixed, and not pure, strategies.

monotonicity Given a bargaining solution for set S, if S is a subset of a larger set W, then the solution outcome in W will dominate the solution for S. In other words, if the players are bargaining over a larger pie, monotonicity ensures that neither one will do worse.

moral hazard A particular condition that provides incentive for an individual to behave differently than he or she normally would. Typically the deviation in behavior is contrary to societal norms.

Nash equilibrium When all players use their best reply strategies in a nonzero sum game.

nature In game theory, random events beyond the control of players.

neural substrate The structures of the brain believed to be linked to specific behaviors or psychological states.

neuroeconomics The scientific study of how economic decision-making is correlated with particular brain activity.

neuromarketing   Branch of marketing that focuses on the brain's responses to stimuli such as advertisements and brand information.

nice   In computer experiments, programs that start out by cooperating.

nonzero sum game   A two-player game in which the pairs of numerical outcomes do not all add up to zero or any other fixed number.

normal form   A way of depicting a game in tabular form. One player selects a row strategy while the second player selects a column strategy.

normative   When results are based on certain principles or axioms; game theory in its purest form is normative, meaning that it prescribes how players should play but does not describe how they actually behave.

outcome   The final result of a game, found when each player has selected a strategy. Outcomes are typically numerical and indicate monetary payoffs or utilities.

Pareto optimal   A solution in which there is no other feasible solution that is at least as good for all players and strictly better for one of them. A Pareto optimal solution is such that any other solution, by comparison, leaves at least one of the players worse off.

payoff   *See* outcome.

payoff matrix   The table of outcomes when games are expressed in strategic form.

perfect information   Games in which the players know all of the previous moves of the game at all times.

player   In a game, a person or agent that selects strategies.

pooling equilibrium   In games of incomplete information, a situation in which players of more than one type all end up playing the same strategy in an equilibrium solution, and are thus indistinguishable.

posterior probability   An updated probability found with the aid of newly gathered information.

power index   In a voting game, a measure of each player's relative strength, influence, or control.

prior probability   A probability estimate that a player starts with.

Prisoner's Dilemma A two-person, nonzero sum game in which each player cooperates or defects. Due to the way the payoffs are defined, the Prisoner's Dilemma has a pure strategy equilibrium in which both players defect. However, the outcomes in this equilibrium are worse than when both players cooperate.

private information In games of incomplete information, payoffs that are known to one player but not to the other players.

probability The likelihood of an outcome's occurrence. This likelihood is expressed by a number between 0 and 1.

probability distribution Given a finite set of distinct outcomes, an assignment of probabilities to the outcomes such that the probabilities sum to 1.

proportional A solution that allocates to each player a share of the overall benefits equal to that player's relative claim or power. In fair division problems, a division method for n players is proportional if it guarantees that each player believes that they receive at least $1/n$ of the total prize.

pure strategy When a player selects only a single action (or a particular sequence of actions) from a set of alternatives.

pure strategy equilibrium An equilibrium in which all players are using a pure, or single, strategy.

random play When players employ random numbers selected from a certain distribution in order to carry out a mixed strategy.

rationality Basis for decision-making that compares different alternatives and selects the best one according to a well-defined utility optimizing decision criterion.

reciprocal altruism A theory that relates altruistic behavior to the likelihood that the behavior will be reciprocated at some point in the future.

reputation The probability that a player is of a certain type or will play a certain strategy.

revelation principle Given a mechanism and any Bayesian-Nash equilibrium of that mechanism, there is a truth-telling equilibrium that yields the same exact outcome.

risk-averse A decision maker who always prefers a certain payoff to a lottery that has an expected value that is equal to that payoff.

risk-seeking A decision maker who always prefers a lottery to a certain payoff equal to the expected value of the lottery.

saddle point A pure strategy equilibrium in a zero sum game in which one player uses a maximin strategy and the opponent uses a minimax strategy.

sealed-bid auction Auction in which each bidder writes down his or her single bid privately and seals it in an envelope to be given to the auction manager.

second-price auction A sealed-bid auction in which the highest bidder wins but pays only the second-highest price bid; also called a Vickrey auction.

sensitivity The probability that a test correctly identifies the presence of a certain condition.

separating equilibrium In games of incomplete information, situation in which players of different types play different strategies in an equilibrium solution, and they are therefore distinguishable from one another.

shading a bid Bidding less than one's full valuation for an item.

Shapley value An efficient solution for cooperative games that allocates to each player his or her average marginal contribution to all coalitions he or she can join.

side payments Money changing hands when costs or profits are shared in cooperative games.

signaling game A game of asymmetric information in which the first player's action sends a message to the second player. The first player knows his or her own type. The second player does not know the first player's type but will act after receiving the message.

simple average An average in which each number is counted only once.

simple game A cooperative game in which the characteristic function equals 0 or 1 for every coalition.

social trap A situation in which individually rational decisions by all members of a group result in an outcome that is not efficient for the group.

solution A methodology by which an outcome is reached; also, the particular strategies that are played in a game and the associated payoffs under those strategies.

specificity The probability that a test correctly identifies the absence of a certain condition.

Stag Hunt A two-person, nonzero sum game in which the players have two strategies: to play together as a team or to play as individuals. The team outcome is superior but each player has to worry about whether the other will act individually.

state of nature One of a number of random outcomes that cannot be controlled by any player.

strategic form *See* normal form.

strategy An action, or series of actions, that a player can undertake.

strategyproof Mechanisms for games of incomplete information in which truthtelling, or revelation of actual valuations, is a dominant strategy for all players.

subgame A well-defined game within a game.

subgame perfect equilibrium A set of equilibrium strategies that provide a Nash equilibrium in every subgame of the game.

sucker's payoff In the Prisoner's Dilemma, the payoff a player receives when they cooperate but the opponent defects.

supergame A game with repeated, identical plays, where each play is itself a well-defined game. The game that is repeated is typically one of the standard two-person games (such as Prisoner's Dilemma).

symmetry Situation in which any outcome that one player can obtain is also obtainable by the other player. It also refers to situations in which equilibrium strategies for the players are identical.

temptation In the Prisoner's Dilemma, the payoff a player receives when they defect but the opponent cooperates.

Tragedy of the Commons A multiplayer game with a limited resource available to all. If all players maximize their use of the resource, it will get permanently depleted.

transitivity With three alternatives A, B, and C, if A is preferred to B, and B is preferred to C, then A must be preferred to C. When transitivity is violated, the preferences are called *intransitive*.

Two Finger Morra A two-player game in which each player simultaneously throws out one or two fingers. One player wins when the sum is even; the other player wins when the sum is odd.

type In games of incomplete information, a player's private information.

Ultimatum game   Game involving two players and a sum of money. One player makes an offer to share some of the money; the other player accepts or rejects that sum. If the offer is rejected, neither player gets anything.

utility   A measure of one's relative gain, pleasure, or satisfaction from a certain outcome.

veto player   A player in a simple game who is a member of all winning coalitions.

Vickrey-Clarke-Groves (VCG) mechanism   An efficient and strategyproof way to allocate multiple goods to multiple players, which charges the players amounts for the goods that are independent of the bids they make.

Von Neumann-Morgenstern utility   A player's preferences over a set of outcomes that meet the following conditions: they are complete, transitive, satisfy the sure thing principle and continuity, and a utility function exists such that $u(A) \geq u(B)$ if and only if outcome A is preferred to outcome B.

voting game   Any number of situations in which voters collectively decide on a choice from a number of alternatives. One standard voting game considers voters as players in a cooperative game.

weighted average   An average in which each number is multiplied by its frequency in the set.

winner's curse   Occurs when the winning bidder discovers that the item they won is not actually worth as much as they paid for it.

zero sum game   A two-player game in which every possible pair of numerical outcomes adds up to zero. In zero sum games, what one player gains, the other loses.

Sources, Resources, and Further Reading

Appendix B

Game theory is fascinating stuff and, as you now know, it has a very mathematical backbone to it and a more psychological side as well. In this appendix I include some suggestions for further reading that encompass both the math and the psychology. I also tell you all of the sources that I have used, and finally, I point you to some websites that are useful.

Further Reading

The Next Step

For those of you who want to pursue the ideas in this book at a higher mathematical level, the following three books do an excellent job of going further into the theory without requiring an advanced degree in mathematics.

Binmore, Ken. *Fun and Games.* Lexington, MA: D.C. Heath, 1991. (This book was updated in 2007 under the title *Playing For Real: A Text on Game Theory.*)

Dutta, Prajit. *Strategies and Games.* Cambridge, MA: MIT Press, 1999.

Gardner, Roy. *Games for Business and Economics, Second edition.* New York: John Wiley & Sons, Inc., 2003.

The Big Leagues

If you're a grad student or researcher and you want to tangle with the deepest and most advanced texts out there, try these. But I warn you, these two books are uncompromisingly written at the highest level.

Fudenberg, Drew, and Jean Tirole. *Game Theory*. Cambridge, MA: MIT Press, 1991.

Myerson, Roger. *Game Theory: Analysis of Conflict*. Cambridge, MA: Harvard University Press, 1997.

More Specialized Sources

The following is a list of other, more specialized sources, all of which I have used in writing this book.

Ariely, Dan. *Predictably Irrational*. New York: HarperCollins, 2008. A fun and very readable collection of ingenious and revealing experiments.

Axelrod, Robert. *The Evolution of Cooperation, revised edition*. New York: Basic Books, 2006. Computer experiments that might yet save the world.

Brams, Steven J. and Alan D. Taylor. *Fair Division: From Cake-Cutting to Dispute Resolution*. Cambridge, UK: Cambridge University Press, 1996. *The* book on fair division.

Brams, Steven J. and Peter C. Fishburn. *Approval Voting, 2nd edition*. New York: Springer, 2007. Everything you need to know about approval voting.

Brandenburger, Adam M. and Barry J. Nalebuff. *Co-opetition*. New York: Currency/Doubleday, 1996. Insights on how to make cooperation work in the competitive world of business.

Camerer, Colin. *Behavioral Game Theory*. Princeton, NJ: Russell Sage Foundation and Princeton University Press, 2003. Indispensible and thorough guide to behavioral game theory.

Damasio, Antonio. *Descartes' Error: Emotion, Reason, and the Human Brain*. New York: Penguin, 1994. Neuroscience and what it means for us by a notable researcher in the field.

Dixit, Avinash, and Barry Nalebuff. *Thinking Strategically: The Competitive Edge in Business, Politics, and Everyday Life*. New York: W. W. Norton & Company, 1991. Some great stories that weave a game-theoretic tapestry.

Frank, Robert H. *Passions Within Reason: The Strategic Role of the Emotions*. New York: W. W. Norton & Company, 1988. Pathbreaking work on the role of our emotions in strategic interaction.

Gigerenzer, Gerd. *Gut Feelings: The Intelligence of the Unconscious.* New York: Viking, 2007. A wise book, full of insights that show that our irrationalities are not all that irrational.

Loewenstein, George, Daniel Read, and Roy F. Baumeister, eds. *Time and Decision: Economic and Psychological Perspectives on Intertemporal Choice.* New York: Russell Sage Foundation, 2003. State-of-the-art contributions that delve into some fascinating topics.

McMillan, John. *Games, Strategies, & Managers.* New York: Oxford University Press, 1992. Brings high-powered theory into the business world, but spares the serious math.

Poundstone, William. *Prisoner's Dilemma.* New York: Doubleday, 1992. Interesting historical material on the War, the Bomb, von Neumann, and game theory, including the PD.

Raiffa, Howard. *The Art and Science of Negotiation.* Cambridge, MA: Belknap/Harvard, 1982. An older but extremely well-written book.

Restak, Richard, M.D. *The Naked Brain: How the Emerging Neurosociety is Changing How We Live, Work, and Love.* New York: Harmony Books, 2006. An easy-to-digest survey of some recent developments in neuroscience.

Ridley, Matt. *The Origins of Virtue: Human Instincts and the Evolution of Cooperation.* New York: Penguin, 1996. Highly readable science writing that sets our place in the animal kingdom.

Smith, Maynard John. *Evolution and the Theory of Games.* Cambridge, UK: Cambridge University Press, 1982. For those super-interested in biology.

Young, H. Peyton. *Equity in Theory and Practice.* Princeton, NJ: Princeton University Press, 1995. Specialized but readable survey of quirky topics in voting, cost sharing, and fairness.

Sources

Chapter 2

Product liability decision tree: adapted from www.litigationrisk.com/Reading%20a%20Tree.pdf.

Chapter 3

War game: Haywood, O. 1954. Military decision and game theory. *Operations Research.* 2,3:39–48.

Chapter 5

Joint ventures as a PD: Parkhe, A., E. Rosenthal, and R. Chandran. 1993. Prisoner's dilemma payoff structure in interfirm strategic alliances: An empirical test. *Omega International Journal of Management Science.* 21,5:531–539.

Advertising wars: Gardner, Roy. *Games for Business and Economics, Second edition.* New York: John Wiley, & Sons, 2003.

Chicken (cars edging forward): Binmore, Ken. *Fun and Games.* Lexington, MA: D.C. Heath, 1991.

Strong and weak PDs: Rapoport, Anatol. 1988. Experiments with N-person social traps I: Prisoner's dilemma, weak prisoner's dilemma, volunteer's dilemma and largest number. *Journal of Conflict Resolution.* 32,3:457–472.

Chapter 6

Fool's Gold: adapted from Gardner, Roy. *Games for Business and Economics, Second edition.* New York: John Wiley, & Sons, Inc., 2003.

Solving Upstart vs. Goliath: Kreps, D. and R. Wilson. 1982. Reputation and imperfect information. *Journal of Economic Theory.* 27:253–279.

Chapter 7

Signaling in the job market: McMillan, John. *Games, Strategies, & Managers.* New York: Oxford University Press, Inc., 1992.

Lemons: Dutta, Prajit. *Strategies and Games: Theory and Practice.* Cambridge, MA: MIT Press, 1999.

Akerlof, G. 1970. The Market for "Lemons": Quality uncertainty and the market mechanism. *The Quarterly Journal of Economics.* 84,3:488–500.

The handicap principle: Zahavi, A. 1975. Mate selection—a selection for a handicap. *Journal of Theoretical Biology.* 53,1:205–214.

Grafen, A. 1990. Biological signals as handicaps. *Journal of Theoretical Biology.* 144:517–546.

Chapter 8

Dividends and future earnings: Benartzi, S., R. Michaely, and R. Thaler. 1996. "Do changes in dividends signal the future or the past?" *The Journal of Finance.* 52,3:1007–1034.

DeAngelo, H., L. DeAngelo, and D. Skinner. 1996. "Reversal of fortune dividend signaling and the disappearance of sustained earnings growth." *Journal of Financial Economics.* 40,3:341–371.

Brinkmanship: Dixit, Avinash and Barry Nalebuff. 2009. *Thinking Strategically: The Competitive Edge in Business, Politics, and Everyday Life.* New York: W. W. Norton & Company, 1991.

Broder, John M. and James Kanter. "China and U.S. hit strident impasse at climate talks," *The New York Times.* December 15, 2009.

Better to give than to receive: "The path to happiness: it is better to give than receive," James Randerson, *The Guardian.* www.guardian.co.uk/science/2008/mar/21/medicalresearch.usa.

Bearing outsized gifts: Ridley, Matt. *The Origins of Virtue.* New York: Penguin, 1996.

Losing your cool and the strategic role of emotions: Frank, Robert. *Passions Within Reason: The Strategic Roles of the Emotions.* New York: W. W. Norton & Company, 1988.

Chapter 10

Bargaining and monotonicity: Kalai, E. and M. Smorodinsky. 1975. Other solutions to Nash's bargaining problem. *Econometrica.* 43,3:513–518.

Gardner, Roy. *Games for Business and Economics, Second edition.* New York: John Wiley & Sons, Inc., 2003.

Chapter 11

Fair division techniques: Steinhaus, H. 1948. The problem of fair division. *Econometrica.* 16:101–104.

Brams, Steven J. and Alan D. Taylor, *Fair Division: From Cake-Cutting to Dispute Resolution.* Cambridge, UK: Cambridge University Press, 1996.

Raiffa, Howard, *The Art and Science of Negotiation.* Cambridge, MA: Harvard/Belknap, 1982.

Chapter 13

Alabama Paradox: Young, H. Peyton. *Equity In Theory and Practice.* Princeton, NJ: Princeton University Press, 1995.

U.S. electoral game: Owen, Guillermo. *Game Theory, Second edition.* New York: Academic Press, 1982.

Approval voting: Brams, Steven, and Peter Fishburn. *Approval Voting, 2nd edition.* New York: Springer, 2007.

Impossibility theorems: Arrow, K. *Social Choice and Individual Values.* New Haven, CT: Yale University Press, 1963.

Gibbard, A. 1973. Manipulation of voting schemes: A general result. *Econometrica.* 41,4:587–601.

Satterthwaite, M. 1975. Strategy-proofness and Arrow's conditions: Existence and correspondence theorems for voting procedures and social welfare functions. *Journal of Economic Theory.* 10,2:187–217.

Chapter 14

Volunteer's Dilemma: Rapoport, Anatol. 1988. Experiments with N-person social traps. *Journal of Conflict Resolution.* 32: 457–472.

Actual behavior in Commons situations: Ostrom, E. 1999. Coping with tragedies of the commons. *Annual Review of Political Science.* 2:493–535.

Ostrom, E. *Governing the Commons: The Evolution of Institutions for Collective Action.* Cambridge, UK: Cambridge University Press, 1990.

Chapter 15

Anglo-Dutch auction: Klemperer, Paul. *Auctions: Theory and Practice.* Princeton, NJ: Princeton University Press, 2004.

Awarding TV rights: McMillan, *Games, Strategies and Managers.* New York: Oxford University Press, 1992.

FCC spectrum auctions: Cave, M., S. Majumdar, and I. Vogelsang, eds. "Spectrum auctions," in *Handbook of Telecommunication Economics.* Chapter 14, pp. 605–639. 2002.

wireless.fcc.gov/auctions/default.htm?job=about_auctions&page=2, and www.pcmag.com/article2/0,2817,2277146,00.asp.

3G wireless auctions: Binmore, K. and P. Klemperer. 2002. The biggest auction ever: The sale of the British 3G telecom licenses. *The Economic Journal.* 112,478:C74–C96.

Chapter 16

A mechanism for cost sharing: Rosenthal, E. 1998. Information and strategyproofness in joint project selection. *Journal of Public Economics.* 68,2:207–221.

Deb, R. and L. Razzolini. 1999. Auction-like mechanisms for pricing excludable public goods. *Journal of Economic Theory.* 88,2: 340–368.

VCG mechanisms: Vickrey, W. 1961. Counterspeculation, auctions, and competitive sealed tenders. *The Journal of Finance.* 16,1:8–37.

Clarke, E. 1971. Multipart pricing of public goods, *Public Choice.* 11:17–33; and Groves, T. 1973. "Incentives in teams," *Econometrica.* 41,4:617–641.

Green, J. and Laffont, J. 1977. Characterization of satisfactory mechanisms for the revelation of preferences for public goods. *Econometrica.* 45,2:427–438.

Generalized second-price auctions: Edelman, B., M. Ostrovsky, and M. Schwarz. 2007. Internet advertising and the generalized second-price auction: Selling billions of dollars worth of keywords. *The American Economic Review.* 97,1:242–259.

Also see services.google.com/awp/en_us/breeze/3004832/index.html.

The revelation principle: Myerson, R. 1979. Incentive compatibility and the bargaining problem. *Econometrica.* 47,1:61–73.

Myerson, R. 1981. Optimal auction design. *Mathematics of Operations Research.* 6, 1:58–73.

Chapter 17

Hawks, Doves, and ESSs: Smith, Maynard. *Evolution and the Theory of Games.* Cambridge, UK: Cambridge University Press, 1982.

Smith, J. Maynard, and G. Price. November 2, 1973. The Logic of Animal Conflict. *Nature.* 246:15–18.

The commitment problem: Frank, Robert H. *Passions within Reason.*

Reciprocal altruism: Trivers, Robert H. 1971. "The evolution of reciprocal altruism." *Quarterly Review of Biology.* 46:35–57.

Emotions are necessary for rational thought: Damasio, Antonio. *Descartes' Error: Emotion, Reason, and the Human Brain.* New York: Penguin, 1994.

Neuromarketing study, Pepsi vs. Coke: McClure, S., J. Li, D. Tomlin, K. Cypert, L. Montague, and P. Montague. 2004. Neural correlates of behavioral preference for culturally familiar drinks. *Neuron.* 44,2:379–387.

The ultimatum game: Güth, W., R. Schmittberger, and B. Schwarze. 1982. An experimental analysis of ultimatum bargaining. *Journal of Economic Behavior and Organization.* 3,4:367–388.

Ultimatum game with fMRI scan: Restak, Richard. *The Naked Brain.* New York: Harmony Books, 2006.

Oxytocin and trust: Restak, Richard. *The Naked Brain.* New York: Harmony Books, 2006

Zak, P., R. Kurzban, and W. Matzner. 2005. Oxytocin is associated with human trustworthiness. *Hormones and Behavior.* 48:522–527.

Kosfeld, M., M. Heinrichs, P. Zak, U. Fischbacher, and E. Fehr. June 2, 2005. Oxytocin increases trust in humans. *Nature.* 435:673–677.

Chapter 18

Prospect theory: Kahneman, D. and A. Tversky. 1979. Prospect theory: An analysis of decision under risk. *Econometrica.* 47,2:263–292.

Chapter 19

Anchoring experiments: Ariely, Dan. *Predictably Irrational.* New York: HarperCollins, 2008.

Physicians and referrals: Bernstein, Peter L. *Against the Gods: The Remarkable Story of Risk.* New York: John Wiley and Sons, Inc., 1996.

Redelmeier, D. and E. Shafir. 1995. Medical decision making in situations that offer multiple alternatives. *Journal of the American Medical Association.* 273,4:302–305.

Intertemporal choice experiment: Thaler, R. 1981. Some empirical evidence on dynamic inconsistency. *Economics Letters* 8:201–207.

Endowment effect: Kahneman, D., J. Knetsch, and R. Thaler. 1991. Anomalies: The Endowment Effect, Loss Aversion, and Status Quo Bias. *The Journal of Economic Perspectives.* 5,1:193–206.

Duke basketball tickets: Ariely, Dan. *Predictably Irrational.*

Chapter 20

Dictator game: Camerer, Colin. *Behavioral Game Theory.* Princeton, NJ: Russell Sage Foundation and Princeton University Press, 2003.

Kahneman, D., J. Knetsch, and R. Thaler. 1986. Fairness and the assumptions of economics. *The Journal of Business.* 59,4:S285–S300.

Frey, B., and I. Bohnet. 1995. Institutions affect fairness: Experimental investigations, *Journal of Institutional and Theoretical Economics.* 151,2:286–303.

Dictator experiment with private booth: Hoffman, E., K. McCabe, K. Shachat, and V. Smith. 1994. "Preferences, property rights and anonymity in bargaining games." *Games and Economic Behavior.* 7:346–380.

Trust in stages: Camerer, Colin. *Behavioral Game Theory.* Princeton, NJ: Russell Sage Foundation and Princeton University Press, 2003.

Berg, J., J. Dickhaut, and K. McCabe. 1995. Trust, reciprocity, and social history. *Games and Economic Behavior.* 10:122–142.

Four-round Centipede game: Camerer, Colin. *Behavioral Game Theory.* Princeton, NJ: Russell Sage Foundation and Princeton University Press, 2003.

McKelvey, R. and T. Palfrey. 1992. An experimental study of the centipede game. *Econometrica.* 60:803–836.

Ultimatum game, baseline results: Camerer, Colin. *Behavioral Game Theory.* Princeton, NJ: Russell Sage Foundation and Princeton University Press, 2003.

More Ultimatum results: Camerer, Colin. *Behavioral Game Theory.* Princeton, NJ: Russell Sage Foundation and Princeton University Press, 2003.

Cameron, L. 1999. Raising the stakes in the ultimatum game: Experimental evidence from Indonesia. *Economic Inquiry.* 27:47–59.

Hoffman, E., K. McCabe, and V. Smith. 1996. On expectations and monetary stakes in ultimatum games. *International Journal of Game Theory.* 25:289–301.

Ultimatum studies by anthropologists: Camerer, Colin. *Behavioral Game Theory.* Princeton, NJ: Russell Sage Foundation and Princeton University Press, 2003.

Age and height effects: Camerer, Colin. *Behavioral Game Theory.* Princeton, NJ: Russell Sage Foundation and Princeton University Press, 2003.

Beauty effect: Schweitzer, M. and S. Solnick. 1999. The influence of physical attractiveness and gender on ultimatum game decisions, *Organizational Behavior and Human Decision Processes.* 79:199–215.

Coordination game experiment: Camerer, Colin. *Behavioral Game Theory.* Princeton, NJ: Russell Sage Foundation and Princeton University Press, 2003.

Friedman, J., ed. 1994. *Problems of Coordination in Economic Activity.* Boston: Kluwer.

Stag Hunt experiment: Cooper, R., D. DeJong, B. Forsythe, and T. Ross. 1990. Selection criteria in coordination games: Some experimental results. *American Economic Review.* 80: 218–233.

Chapter 21

Dominant strategies in practice: Camerer, Colin. *Behavioral Game Theory.* Princeton, NJ: Russell Sage Foundation and Princeton University Press, 2003.

Beard, T. and R. Beil. 1994. Do people rely on the self-interested maximization of others? An experimental test. *Management Science.* 40:252–262.

Ducks play Nash equilibrium: Harper, D. 1982. Competitive foraging in mallards: "ideal free" ducks. *Animal Behavior.* 30:575–584.

Penalty kicks: Camerer, Colin. *Behavioral Game Theory.* Princeton, NJ: Russell Sage Foundation and Princeton University Press, 2003.

Palacios-Huerta, I. 2003. Professionals play minimax. *The Review of Economic Studies.* 70,2:395–415.

Chiappori, P., S. Levitt, and T. Groseclose. 2002. Testing mixed strategy equilibria when players are heterogeneous: The case of penalty kicks in soccer. *American Economic Review.* 92,4:1138–1151.

Bar-Eli, M. O. Azar, I. Ritov, Y. Keidar-Levin, and G. Schein. 2007. Action bias among elite soccer goalkeepers: The case of penalty kicks. *Journal of Economic Psychology.* 28,5:606–621.

Beauty contests: Camerer, Colin. *Behavioral Game Theory.* Princeton, NJ: Russell Sage Foundation and Princeton University Press, 2003.

Nagel, R. 1995. Unravelling in guessing games: An experimental study. *American Economic Review.* 85:1313–1326.

First PD experiment: Poundstone, William. *Prisoner's Dilemma.* New York: Doubleday, 1992.

PD subjects cooperative/competitive: Kelley, H. and A. Stahelski. 1970. Social interaction basis of cooperators' and competitors' beliefs about others. *Journal of Personality and Social Psychology.* 16:66–91.

PD with preplay interaction: Frank, R., T. Gilovich, and D. Regan. 1993. The evolution of one-shot cooperation: An experiment. *Ethology and Sociobiology* 14,4:247–256.

Heuristics and unconscious intelligence: Gigerenzer, Gerd. *Gut Feelings.* New York: Viking, 2007.

Chapter 22

The Chain Store Paradox: Selten, R. 1978. The chain store paradox. *Theory and Decision.* 9:127–159.

Gang of four PD paper: D. Kreps, D., P. Milgrom, J. Roberts, and R. Wilson. 1982. Rational cooperation in the finitely repeated prisoner's dilemma. *Journal of Economic Theory.* 27,2:245–252.

PD experiment: J. Andreoni, J. and J. Miller. 1993. Rational cooperation in the finitely repeated prisoner's dilemma: Experimental evidence. *The Economic Journal.* 103:570–585.

Axelrod's experiment: Axelrod, Robert. *The Evolution of Cooperation.* New York: Basic Books, 1984.

Websites

As we all know, tons of stuff on any subject can be found online. This is true for game theory as well. At the same time, we have to understand that once in a while, a website disappears, URL's change, people stop supporting some pages, and what's here one day is gone the next.

I use Wikipedia for all manner of things, although I'm always wary of what I find. I have looked at a lot of game theory material on Wikipedia and I find some of it useful and relevant (as well as accurate), while some of it is not.

What about the rest of the web? If you want to locate scholarly articles, I highly recommend just going into Google Scholar and typing some names or search terms.

Below I list a few sites of interest that I've come across, in no particular order. You can always explore on your own as well—have a game theory adventure!

General game theory:
gametheory.net

Software to solve games:
www.gambit-project.org/doc/index.html

History/Chronology of game theory: www.econ.canterbury.ac.nz/personal_pages/paul_walker/gt/hist.htm

Banzhaf power index computation: www.math.temple.edu/~cow/bpi.html

VCG auctions, by Jeffrey Ely: cheeptalk.files.wordpress.com/2009/07/vcg.pdf

3G telecom auctions by Binmore and Klemperer: www.nuff.ox.ac.uk/users/klemperer/biggestpaper.pdf

Prisoner's Dilemma, Stag Hunt, and other games: plato.stanford.edu/entries/prisoner-dilemma

Index

A

adjusted winner procedures, 155-157
adverse selection, 220-221
Allais, Maurice, 258
altruism, 243-244, 282-283
ambiguity aversion, 257-258
analyzing
    low threshold Tragedy of the Commons, 196
    Shapley value, 167
    Stag Hunt game, 76
    Vickrey-Clarke-Groves mechanisms, 229
anchoring, 267
    Columbia River example, 267
    Social Security digits, 268
    two-person bargaining, 268-270
Anglo-Dutch auctions, 211
animal kingdom
    altruism, 243
    strategies
        evolutionarily stable strategy (ESS), 240-242
        Hawk-Dove game, 238-240
anticipating other's moves, 293-294
approval voting, 187-188
Ariely, Dan, 268
Arrow, Kenneth, 188-189
ascending-bid auctions, 206, 216
asymmetric information, 86
    adverse selection, 220-221
    auctions, 206

defined, 87
department store jewelry purchases, 91
economic conditions, 90
Fool's Gold game, 87-90
    payoffs, 89
    summary, 88
Internet prescription medications, 91
lemons, 103
moral hazards, 221-222
payoffs, 59-60
auctions
    Anglo-Dutch, 211
    asymmetric information, 206
    common-value, 210
    cost-sharing mechanisms, 222-226
    examples
        FCC spectrum, 214-217
        New Zealand airwaves, 212-213
        TV rights, 213-214
    first-price
        ascending-bid, 206
        sealed bids, 207
        shading bids, 208-209
        strategies, 207
    generalized second-price, 229-231
    honest bidding, 229
    oft-used, 210
    overview, 205
    package bidding, 215
    sealed-bid, 213-214

second-price, 209
  New Zealand airwaves, 212-213
  Vickrey, William, 211-212
shading bids, 208
valuations of items, 206
Vickrey, William, 211-212
Vickrey-Clarke-Groves mechanism
  analyzing, 229
  calculating, 228-229
  overview, 226-228
winner's curse, 210
averages
  expected values, 20
  maximins, 22
  minimax, 22
  probabilities, 19
  simple, 17
  weighted, 18-19
Axelrod, Robert, 311
axiomatic approaches, 130

B

backward induction, 279-281
bankruptcies
  Kalai-Smorodinsky solution, 144-146
  solutions, 140-143
Banzhaf III, John F., 184
Banzhaf Power Index, 184-185
bargaining, 125-127
  setting up, 126
  two-player, 135, 268-270
    conditions, 136
    disagreement points, 135
    equality, 137
    feasibility, 138
    individual rationality, 138
    irrelevant alternatives, 139
    Kalai-Smorodinsky solution, 143-146
    linear transformations, 139
    Nash solution, 140-143

Pareto optimality, 138
setting up, 136
summery, 139
Battle of the Bismarck Sea, 39
Bayes' rule, 29-31
Bayesian games, 32, 94
Bayesian-Nash equilibrium, 94
beauty contests, 214, 293-294
behaviors
  beauty contests, 214, 293-294
  coordination games, 286
  decision-making
    anchoring, 267-270
    bounded rationality, 298-299
    endowment effect, 273-274
    failure of invariance, 267
    frames of reference, 265
    heuristic approaches, 299
    logical fallacies, 299
    mental accounting, 266
    preference patterns for gains and losses over time, 272
    preference patterns for money over time, 270-272
    rationality, 304
    unconscious intelligence, 300
  dominant strategies, 289-291
  mixed strategies, 291
    duck pond, 291
    Nash equilibrium, 56-57
    soccer penalty kicks, 292
    zero sum games, 36, 42-47
  Prisoner's Dilemma, 295
    cigarette advertising example, 70-71
    computer tournament, 312-315
    cooperation versus defecting, 67
    cooperative strategy, 68
    defecting, 68
    emission reduction, 71
    equilibrium, 68
    experiments, 295
    incomplete information, 85-86

joint business venture example, 69-70
overview, 66
payoffs, 67
repeating, 305-306
strategies, 67
strong and weak, 78-79
prospect theory, 259
behavioral decision-making associations, 259
elements, 259
gains versus losses, 259-262, 272
probabilities, 263-264
relative values, 262-263
rationality, 298-299
bounded, 298-299
decision-making, 304
payoffs, 308
Ultimatum game, 281-282
altruism, 282-283
demographic variables, 284
physical attractiveness, 284
raising the stakes, 283
repeating the game, 283
unconscious intelligence, 300
utilities
behavior indicators, 255
Bernoulli's theory, 253
certainty equivalents, 256
contradictions, 257-258
proportionality, 257
risk averse, 256
St. Petersburg Paradox, 251-254
von Neumann's/Morgenstern's theory, 253-255
Bernoulli, 251-253
Binmore, Ken, 72
biology
altruism, 243-244
emotional influences on decision-making, 242-243
neuroscience
emotional power over rationality, 245-246

neuroeconomics, 246-248
neuromarketing, 246-247
trust, 248-249
blocs, 132
blue/red balls probability game, 257-258
Borda, Jean-Charles de, 178
Borda point-count tallies, 179-180
bounded rationality, 298-300
branding, 107
brinkmanship, 115-117
budget balanced, 232
budgeting versus efficiency, 231-233
buyer-seller games
department store jewelry purchases, 91
economic conditions, 90
Fool's Gold, 87
payoffs, 89
summary, 88
Internet prescription medications, 91

C

cake sharing example
Last Diminisher Procedure, 150
three-people, 148-150
two-people, 147-148
calculating
Banzhaf Power Index, 184-185
Shapley value, 166-167
Vickrey-Clarke-Groves mechanisms, 228-229
Camerer, Colin, 278
carbon emission reduction, 71
categories of games
cooperative, 13
characteristic function, 163
defined, 163
establishing players, 163
power generation example, 168-169
setting up, 163
Shapley value, 164-167
solving, 170-173

extensive form, 10-11
incomplete information, 12
nonzero
    Chicken, 72-74, 77
    Coordination, 54
    credibility, 61-63
    Nash equilibrium, 55-60
    Stag Hunt, 75-77
nonzero sum, 8-10, 54
perfect information, 11-12
zero sum
    constant sum, 7-8, 41-42
    defined, 6
    equilibrium solution, 38-39
    misconceptions, 50-51
    setting up, 35
    solutions, 38-39, 42-50
    strategies, 36-38
    Two Finger Morra, 6-7
    wars, 39-40
Centipede games
    behavioral study, 279-280
    overview, 61-62
certainty
    ambiguity aversion, 257-258
    equivalents, 256
Chain Store Paradox, 302
    deterrence, 304
    equilibrium, 303
    rational decision-making, 304
characteristic functions
    cooperative games, 163
    voting, 176
Chicken game, 72-74
    payoffs, 73
    strategies, 73
    win-win transformation, 77
choosing
    decision criterions, 22
    strategic choices, 23-24
cigarette advertising example, 70-71
cleaner fish versus host fish, 243
coalitions, 130

coin flipping game, 251-252
Coke versus Pepsi, 246-247
college mug example, 273
collusion, 234
Columbia River example, 267
commitment, 113-114, 242
common-value auctions, 210
computer tournament for cooperation, 311
    Bayesian updating mechanism, 312
    cooperation in competitive environments, 315
    evolutionary pressures, 314
    future payoffs, 314-315
    nice/not nice programs, 312
    payoffs, 311
    random defection, 312
    strategies, 312
    tit for tat, 312-313
conclusions
    test sensitivity, 30
    trustworthiness, 29
        Bayes' rule, 29-31
        test reliability, 31-32
conditions
    prospect theory, 259
    Shapley value, 164-166
    two-player bargaining, 136
Condorcet cycles, 178-179
Condorcet, Marquis de, 178
conflicts
    cooperation, 162
    Prisoner's Dilemma, 85-86
conjunction fallacy, 299
considering other player's positions, 15
constant sum games, 7-8, 41-42
Constitutional Convention, 282
convex, 261
cooperation
    bargaining, 125-127
        conditions, 136
        disagreement points, 135
        equality, 137
        feasibility, 138

Index

individual rationality, 138
irrelevant alternatives, 139
Kalai-Smorodinsky solution, 143-146
linear transformations, 139
Nash's solution, 140-143
Nash's standard model, 135
Pareto optimality, 138
setting up, 126-136
symmetry, 139
characteristic function, 163
computer tournament, 311
  Bayesian updating mechanism, 312
  cooperation in competitive environments, 315
  evolutionary pressures, 314
  future payoffs, 314-315
  nice/not nice programs, 312
  payoffs, 311
  random defection, 312
  strategies, 312
  tit for tat, 312-313
conflicts, 162
cost-sharing problems, 161-162
  mechanism design, 222-226
  power generation example, 168-169
  Shapley value, 164-167
  solving, 171-173
  taxi sharing fairness, 161-162
defined, 163
establishing players, 163
fair division problems, 127-129
  adjusted winner procedure, 155-157
  Knaster-Steinhaus Procedure, 151-154
  Last Diminisher Procedure, 150
  proportional allocation procedure, 157-158
  sharing, 155
  three-person cut and choose example, 148-150
  two-person cut and choose example, 147-148
folk theorem, 309-310
gang of four's theory, 305-307

power generation example, 168-169
Prisoner's Dilemma
  cigarette advertising example, 70-71
  cooperation versus defecting, 67
  cooperative strategy, 68
  defecting, 68
  emission reduction, 71
  equilibrium, 68
  incomplete information, 85-86
  joint business venture example, 69-70
  overview, 66
  payoffs, 67
  repeating, 308
  strategies, 67
  strong and weak, 78-79
setting up, 163
Shapley value
  analysis, 167
  calculating, 166-167
  conditions, 164-166
social networking, 316
solving
  core method, 171-173
  profit-sharing problems, 170-171
Tragedy of the Commons, 203
voting, 131-133
  blocs, 132
  proportional representation, 132
  third party candidates, 131
  weighted point counting system, 131
coordination games, 8-10, 54, 285
  Nash equilibrium
    asymmetric payoffs, 59-60
    mixed strategy, 56-57
    payoffs, 57-59
    pure strategy, 55
    strategies, 285
  Stag Hunt, 286-287
  theory versus behavior, 286
  two-way communication, 286
core of cooperative games, 171-173
corporate signals, 111-112

cost of education game, 101-102
cost-sharing problems
   mechanism design, 222-226
   power generation example, 168-169
   Shapley value
      analysis, 167
      calculating, 166-167
      conditions, 164-166
   solving, 171-173
   taxi sharing fairness, 161-162
costly commitments, 113-114
credibility
   nonzero games, 61-63
   subgame perfect equilibrium, 64
Cuban missile crisis, 115

D

decision-making
   anchoring, 267
      Columbia River example, 267
      Social Security digits, 268
      two-person bargaining, 268-270
   bounded rationality, 298-299
   criterions, 22, 25
   emotional influences, 242-243
   endowment effect, 273-274
   failure of invariance, 267
   frames of reference, 265
   group
      approval, 187-188
      blocs, 132
      characteristic functions, 176
      as group decision, 175
      impossibility theorem, 188-189
      individual rankings, 179-180
      manipulating, 189
      pairwise comparisons, 178-179
      perfection, 188
      proportionality, 132, 181-186
      swing states, 185
      swing votes, 183
      theories, 178
      third party candidates, 131
      U.S. electoral system, 186-187
      veto players, 176-178
      weighted point counting system, 131
   heuristic approaches, 299
   logical fallacies, 299
   mental accounting, 266
   preference patterns over time
      gains and losses, 272
      money, 270-272
   prospect theory
      gains versus losses, 259-262
      probabilities, 263-264
      relative values, 262-263
   rationality, 304
   theory
      defined, 23
      overbooking strategies, 24-26
      strategic choices, 23-24
   trees, 26-27
   unconscious intelligence, 300
   utilities
      behavior indicators, 255
      Bernoulli's theory, 253
      certainty equivalents, 256
      contradictions, 257-258
      defined, 9
      proportionality, 257
      risk averse, 256
      St. Petersburg Paradox, 251-254
      von Neumann's/Morgenstern's theory, 253-255
deforestation, 194-195
department store jewelry purchases, 91
deterrence, 304
diagrams
   Condorcet cycles, 178-179
   decision trees, 26-27
   extensive form, 10-11
Dictator game, 276-277
diminishing returns, 252

disagreement points, 136
discount rates, 271
dominant strategies, 289-291
Dresher, Melvin, 66
duck pond experiment, 291
Dutch auctions, 210

E

economic conditions, 90
efficiency versus budgeting, 231-233
elderly medical insurance policies, 106
electoral votes, 181, 186-187
eliciting true valuations for group benefit
    auctionlike procedure, 222-226
    generalized second-price auctions, 229-231
    limitations, 231-234
    revelation principle, 233
    strategyproof, 232
    Vickrey-Clarke-Groves, 226-229
Ellsberg, Daniel, 257-258
emission reduction, 71
emotions
    altruism, 243-244
    influence on decision-making, 242-243
    power over rationality, 245-246
    trust, 248-249
        Centipede game, 279-280
        Dictator game, 276-277
        stages, 277-278
endowment effect, 273-274
English auctions, 206
equality, 137
equilibrium
    Bayesian-Nash, 94
    Chain Store Paradox, 303
    mixed strategy, 42
    Nash, 55
        coordination games, 285
        mixed strategy, 56-57, 73

        payoffs, 57-60
        pure strategy, 55
    pooling, 101-102
    Prisoner's Dilemma, 68
    separating, 100
    subgame perfect, 64
    zero sum games, 38-39
equitable, 156
erratic behavior, 117-118
ESS (evolutionarily stable strategy), 240-242
evaluating
    low threshold Tragedy of the Commons, 196
    Shapley value, 167
    Stag Hunt game, 76
    Vickrey-Clarke-Groves mechanisms, 229
evens/odds game, 6-7
expected values, 20
    overbooking strategies, 25
    probabilities, 263-264
extensive form games, 10-11

F

failure of invariance, 267
fair division, 127-129
    adjusted winner procedure, 155-157
    cost-sharing problems
        mechanism design, 222-226
        power generation example, 168-169
        Shapley value, 164-167
        solving, 171-173
        taxi sharing, 161-162
    cut and choose example
        Last Diminisher Procedure, 150
        three-people, 148-150
        two-people, 147-148
    Knaster-Steinhaus Procedure, 151-154
    proportional allocation procedure, 157-158
    sharing, 155

farmer water source problem, 199
FCC spectrum auctions, 214-217
feasibility, 138
first-price auctions
    ascending-bid, 206, 216
    sealed bids, 207
    shading bids, 208-209
    strategies, 207
fishing, 194-195
fitness, 238
Flood, Merrill, 66
fMRI results, 246
folk theorem
    computer cooperation tournament, 314-315
    overview, 309-310
Fool's Gold game, 87-90
    economic conditions, 90
    payoffs, 89
    summary, 88
forestation, 194-195
fractional apportionments, 181
frames of reference, 265
Franklin, Benjamin, 222
Free Rider problem, 201-204
functions
    behavior indicators, 255
    Bernoulli's theory, 253
    certainty equivalents, 256
    contradictions, 257-258
    proportionality, 257
    risk averse, 256
    von Neumann's/Morgenstern's theory, 253-255
future of neuroscience, 249

G

gains and losses, 259-262, 272
games
    Bayesian, 32, 94
    Chicken, 72-74
        payoffs, 73
        strategies, 73
        win-win transformation, 77
    coin flipping, 251-252
    constant sum, 7-8, 41-42
    cooperative, 13, 129-130
        characteristic function, 163
        computer tournament, 311-315
        conflicts, 162
        core, 171-173
        cost-sharing problems. *See* cost-sharing problems
        defined, 163
        establishing players, 163
        fair division. *See* fair division
        folk theorem, 309-310
        gang of four's theory, 305-307
        power generation example, 168-169
        profit-sharing problems, 170-171
        setting up, 163
        Shapley value, 164-167
        social networking, 316
        Tragedy of the Commons, 203
        voting, 131-133
    coordination, 285
        Nash equilibrium. *See* Nash, equilibrium
        Stag Hunt, 286-287
        theory versus behavior, 286
        two-way communication, 286
    defined, 4
    dictator, 276-277
    extensive form, 10-11
    Hawk-Dove, 238-240
    incomplete information, 12
    nonzero
        Chicken, 72-74
        payoffs, 73
        Coordination, 54
        credibility, 61-63
        Nash equilibrium, 55-60
        Stag Hunt, 75-76

nonzero sum, 8-10
    Coordination, 8-10
    Prisoner's Dilemma. *See* Prisoner's Dilemma
    Wright's definition, 54
perfect information, 11-12
Prisoner's Dilemma
    Bayesian updating mechanism, 312
    cigarette advertising example, 70-71
    computer tournament, 312-315
    cooperation, 67-68, 315
    defecting, 68
    emission reduction, 71
    equilibrium, 68
    evolutionary pressures, 314
    experiments, 295
    incomplete information, 85-86
    joint business venture example, 69-70
    overview, 66
    payoffs, 311-315
    random defection, 312
    strategies, 67-68, 312
red/blue balls, 257-258
repeating, 301-302
    deterrence, 304
    equilibrium, 303
    folk theorem, 309-310
    gang of four's cooperation theory, 305-307
    Prisoner's Dilemma, 305-308
    rational decision-making, 304
    supergames, 303
signaling, 98
    branding, 107
    cost of education, 101-102
    handicap principle, 107-108
    job market, 97-100
    nuanced signals. *See* nuanced signals
    pooling equilibrium, 101-102
    quality, 102-106
solving, 13-14

Stag Hunt, 75
    analyzing, 76
    payoffs, 75-76, 286
    win-win transformation, 77
supergames, 303
Ultimatum
    altruism, 282-283
    behavioral study, 281-284
    demographic variables, 284
    neuroscience, 247-248
    physical attractiveness, 284
    raising the stakes, 283
    repeating the game, 283
zero sum
    CBS example solution, 45-47
    constant sum, 7-8, 41-42
    defined, 6
    equilibrium solution, 38-39
    linear programming, 49-50
    misconceptions, 50-51
    mixed strategies, 36, 42
    NBC example solution, 43-44
    pure strategies, 36-38
    saddle points, 48-49
    setting up, 35
    Two Finger Morra, 6-7
    wars, 39-40
gang of four's cooperation theory, 305-307
generalized second-price auctions, 229-231
generosity, 277
Gibbard, Allan, 189
gift giving, 118-119
Gigerenzer, Gerd, 299
glove example, 176-177
Goliath versus Upstart, 302
group decisions, 131-133
    approval, 187-188
    blocs, 132
    characteristic functions, 176
    as group decision, 175
    impossibility theorem, 188-189

individual rankings, 179-180
manipulating, 189
pairwise comparisons, 178-179
perfection, 188
proportionality
   power, 182-186
   problems, 181-182
   representation, 132
swing states, 185
swing votes, 183
theories, 178
third party candidates, 131
U.S. electoral system, 186-187
veto players, 176-178
weighted point counting system, 13
group versus individual strategies
   Free Rider, 201-204
   Tragedy of the Commons
      Hardin's description, 193-194
      high thresholds, 197-198
      low thresholds, 194-197
      moral hazards, 199
      overfishing/deforestation, 194-195
      real-world cooperation, 203
      regulating versus privatizing, 203
      solving, 199, 203
   Volunteer's Dilemma, 200-203

H

Hamilton, Alexander's apportionment rule, 181
handicap principle, 107-108
Hardin, Garrett, 193
Hawk-Dove game, 238-240
heuristic procedures, 298-299
high threshold Tragedy of the Commons, 197-198
history, 3-4
HIV tests, 31
homemade priors, 306

honesty
   adverse selection, 220-221
   mechanism designs, 222
      calculating, 233
      collusion, 234
      cost-sharing, 222-226
      efficiency versus budgeting, 231-233
      generalized second-price auctions, 229-231
      revelation principle, 233
      strategyproof, 232
      Vickrey-Clarke-Groves, 226-229
   moral hazards, 221-222
horse market profit-sharing situation, 170-171
host fish versus cleaner fish, 243
hostages and enforcement, 79
Hume, David, 75
hyperbolic discounting, 271

I

imperfect information, 84-85
impossibility theorem for voting, 188-189
incomplete information, 12, 85-86
   Prisoner's Dilemma example, 85-86
   Upstart versus Goliath example, 92-94
individual versus group strategies
   Free Rider, 201-204
   Tragedy of the Commons
      Hardin's description, 193-194
      high thresholds, 197-198
      low thresholds, 194-197
      moral hazards, 199
      overfishing/deforestation, 194-195
      real-world cooperation, 203
      regulating versus privatizing, 203
      solving, 199, 203
   Volunteer's Dilemma, 200-203
individuals
   anticipating other's moves, 293-294
   considering other player's positions, 15

cooperation, 163
  bargaining. *See* bargaining
  characteristic function, 163
  computer tournament, 311-315
  conflicts, 162
  cost-sharing problems. *See* cost-sharing problems
  defined, 163
  establishing players, 163
  fair division. *See* fair division
  folk theorem, 309-310
  gang of four's theory, 305-307
  power generation example, 168-169
  Prisoner's Dilemma, 67-71, 78-79, 85-56
  setting up, 163
  Shapley value, 164-167
  social networking, 316
  solving, 17-173
  Tragedy of the Commons, 203
  voting, 131-133
defined, 4
honesty
  adverse selection, 220-221
  mechanism designs. *See* mechanism designs
  moral hazards, 221-222
losing your temper, 119-120
power
  emotions over rationality, 245-246
  generation example, 168-169
  indices, 183-186
  veto players, 176-178
  voting proportionality, 182-186
rankings, 179-180
rationality, 138
reputations, 112
temptations
  adverse selection, 220-221
  mechanism designs. *See* mechanism designs
  moral hazards, 221-222

trust, 248-249
  Centipede game, 279-280
  conclusions, 29-32
  Dictator game, 276-277
  stages, 277-278
  veto, 176-178
information
  asymmetric, 86
    adverse selection, 220-221
    defined, 87
    department store jewelry purchases, 91
    economic conditions, 90
    Fool's Gold game, 87-90
    Internet prescription medications, 91
    lemons, 103
    moral hazards, 221-222
  imperfect, 84-85
  incomplete, 85-86
    Prisoner's Dilemma example, 85-86
    Upstart versus Goliath example, 92-94
  perfect, 83-85
Internet
  access example, 130
  advertising, 229-231
  prescription medication purchases, 91
intransitive, 179
irrationality, 308
irrelevant alternatives, 139, 269
iterated dominance, 294

J

job market game
  analysis, 100
  cost of education, 101-102
  data summary, 100
  overview, 97-99
joint business venture Prisoner's Dilemma example, 69-70

K

Kahneman, Daniel
    failure of invariance, 267
    mental accounting, 266
    prospect theory, 259
Kalai-Smorodinsky solution, 143-146
Kanster-Steinhaus Procedure, 151-154
Klemperer, Paul, 211

L

Last Diminisher Procedure, 150
lemon game, 102
life and death outcomes, 266
limited cognitive abilities, 298-299
linear programming, 49-50
linear transformations, 139
logical fallacies, 299
lose-lose outcome transformations, 77
    Chicken game, 77
    hostages and enforcement, 79
    Prisoner's Dilemma, 78-79
    Stag Hunt game, 77
losing your temper, 119-120
loss aversion, 261
losses versus gains, 259-262, 272
lotteries, 254
low threshold Tragedy of the Commons, 194-197
    analyzing, 196
    payoffs, 197
    three-person, 196
    two players, 195

M

making decisions
    anchoring, 267
        Columbia River example, 267
        Social Security digits, 268
        two-person bargaining, 268-270
    bounded rationality, 298-299
    criterions, 22, 25
    emotional influences, 242-243
    endowment effect, 273-274
    failure of invariance, 267
    frames of reference, 265
    group
        approval, 187-188
        blocs, 132
        characteristic functions, 176
        as group decision, 175
        impossibility theorem, 188-189
        individual rankings, 179-180
        manipulating, 189
        pairwise comparisons, 178-179
        perfection, 188
        proportionality, 132, 181-186
        swing states, 185
        swing votes, 183
        theories, 178
        third party candidates, 131
        U.S. electoral system, 186-187
        veto players, 176-178
        weighted point counting system, 131
    heuristic approaches, 299
    logical fallacies, 299
    mental accounting, 266
    preference patterns over time
        gains and losses, 272
        money, 270-272
    prospect theory
        gains versus losses, 259-262
        probabilities, 263-264
        relative values, 262-263
    rationality, 304

Index   353

theory
  defined, 23
  overbooking strategies, 24-26
  strategic choices, 23-24
trees, 26-27
unconscious intelligence, 300
utilities
  behavior indicators, 255
  Bernoulli's theory, 253
  certainty equivalents, 256
  contradictions, 257-258
  defined, 9
  proportionality, 257
  risk averse, 256
  St. Petersburg Paradox, 251-254
  von Neumann's/Morgenstern's theory, 253-255
manipulating voting systems, 189
marginal costs, 166
marginal values, 166
market
  entry games, 62-63
  failures, 90
  successes, 90
Mason, George, 282
mathematics
  averages
    expected values, 20
    maximins, 22
    minimax, 22
    probabilities, 19
    simple, 17
    weighted, 18-19
  calculating
    Banzhaf Power Index, 184-185
    Shapley value, 166-167
    Vickrey-Clarke-Groves mechanisms, 228-229
  models, 5
  probabilities, 19
    expected values, 263-264
    prior, 30-31

prospect theory, 263-264
red/blue balls probability game, 257-258
theories versus real life, 5
weighted point counting systems, 131
maximins, 22, 25
Maynard Smith, John, 238
  evolutionarily stable strategy (ESS), 240-242
  Hawk-Dove game, 238-240
McMillan, John, 213-214
mechanism designs, 222
  cost-sharing, 222-226
  generalized second-price auctions, 229-231
  limitations
    calculating, 233
    collusion, 234
    efficiency versus budgeting, 231-233
  revelation principle, 233
  strategyproof, 232
  Vickrey-Clarke-Groves
    analyzing, 229
    calculating, 228-229
    overview, 226-228
medical insurance for elderly, 106
mental accounting, 266
menu pricing, 268
minimax, 22
mixed Nash equilibrium, 73
mixed strategies, 291
  duck pond, 291
  Nash equilibrium, 56-57
  soccer penalty kicks, 292
  zero sum games, 36, 42
    CBS example solution, 45-47
    NBC example solution, 43-44
monetary bargaining
  Kalai-Smorodinsky, 143-146
  Nash, 140-143
money over time preference patterns, 270-272

monotonicity, 144
moral hazards
    defined, 199
    individual temptations, 221-222
    Tragedy of the Commons, 199
Morgenstern, Oskar, 253-255

N

Nader, Ralph, 131
Nash
    equilibrium, 55
        asymmetric payoffs, 59-60
        mixed strategy, 56-57, 73
        payoffs, 57-59
        pure strategy, 55
        strategies, 285
    two-player bargaining model, 135-136, 140-143
Nash, John, 55
NCAA tickets example, 273
network programming game, 7-8
    CBS, 45-47
    NBC, 43-44
Neumann, John von, 253-255
neural substrate, 247
neuroeconomics, 246-248
neuromarketing, 246-247
neuroscience
    emotional power over rationality, 245-246
    future, 249
    neuroeconomics, 246-248
    neuromarketing, 246-247
    trust, 248-249
New Zealand airwaves auction, 212-213
nice programs, 312
nonreciprocal altruism, 244
nonzero games
    Chicken, 72-74
        payoffs, 73
        strategies, 73
        win-win transformation, 77

Coordination, 54
credibility, 61
    Centipede games, 61-62
    Market Entry games, 62-63
Nash equilibrium, 55
    asymmetric payoffs, 59-60
    mixed strategy, 56-57
    payoffs, 57-59
    pure strategy, 55
Stag Hunt, 75, 286-287
    analyzing, 76
    payoffs, 75-76
    win-win transformation, 77
nonzero sum games, 8-10
    Prisoner's Dilemma
        cigarette advertising example, 70-71
        computer tournament, 312-315
        cooperation, 67-68, 315
        defecting, 68
        emission reduction, 71
        equilibrium, 68
        experiments, 295
        incomplete information, 85-86
        joint business venture example, 69-70
        overview, 66
        payoffs, 67
        repeating, 305-306
        strategies, 67-68, 312
        strong and weak, 78-79
    Wright's definition, 54
not nice programs, 312
nuanced signals
    commitment, 113-114
    defined, 114
    reputations, 112
    spending money, 111-112
    threat strategies, 115
        brinkmanship, 115-117
        erratic behavior, 117-118
    veiled strategies
        gift giving, 118-119
        losing your temper, 119-120

O

odds/evens game, 6-7
oft-used auctions, 210
origins, 3-4
Ostrom, Elinor, 203
outcomes, 4
    life and death, 266
    lose-lose, 77
        Chicken game, 77
        hostages and enforcement, 79
        Prisoner's Dilemma, 78-79
        Stag Hunt game, 77
    win-win, 77
        Chicken game, 77
        hostages and enforcement, 79
        Prisoner's Dilemma, 78-79
        Stag Hunt game, 77
overbooking strategies, 24-26
overfishing, 194-195
Owen, Guillermo, 187
oxytocin, 249

P

package bidding, 215
pairwise comparisons, 178-179
Pareto optimality, 138
Pareto, Vilfredo, 138
partial market successes, 90
payoffs
    asymmetric, 59-60
    certainty equivalents, 256
    Chicken game, 73
    computer tournament for cooperation, 314-315
    defined, 4
    folk theorem, 309-310
    Fool's Gold, 89
    low threshold Tragedy of the Commons, 197
    Nash equilibrium, 57-60
    nonzero games, 73
    preferences for money over time, 270-272
    Prisoner's Dilemma, 67
    rational play, 308
    risk averse players, 256
    Stag Hunt game, 75-76, 286
peacocks, 107-108
penalty kick game, 10
Penrose, L.S., 185
Pepsi versus Coke, 246-247
perfect information games, 11-12, 83-85
Perot, Ross, 131
Philadelphia Contributionship company, 222
physical attractiveness Ultimatum game, 284
physician study, 269
players
    anticipating other's moves, 293-294
    considering other player's positions, 15
    cooperation, 163
        bargaining. *See* bargaining
        characteristic function, 163
        computer tournament, 311-315
        conflicts, 162
        cost-sharing problems. *See* cost-sharing problems
        defined, 163
        establishing players, 163
        fair division. *See* fair division
        folk theorem, 309-310
        gang of four's theory, 305-307
        power generation example, 168-169
        Prisoner's Dilemma, 67-71, 78-79, 85-56
        setting up, 163
        Shapley value, 164-167
        social networking, 316
        solving, 17-173
        Tragedy of the Commons, 203
        voting, 131-133
    defined, 4
    honesty
        adverse selection, 220-221
        mechanism designs. *See* mechanism designs
        moral hazards, 221-222

losing your temper, 119-120
power
    emotions over rationality, 245-246
    generation example, 168-169
    indices, 183-186
    veto players, 176-178
    voting proportionality, 182-186
rankings, 179-180
reputations, 112
temptations
    adverse selection, 220-221
    mechanism designs. *See* mechanism designs
    moral hazards, 221-222
trust, 248-249
    Centipede game, 279-280
    conclusions, 29-32
    Dictator game, 276-277
    stages, 277-278
two-person bargaining
    anchoring, 268-270
    conditions, 136
    disagreement points, 135
    equality, 137
    feasibility, 138
    individual rationality, 138
    irrelevant alternatives, 139
    Kalai-Smorodinsky solution, 143-146
    linear transformations, 139
    Nash solution, 140-143
    Nash's standard model, 135
    Pareto optimality, 138
    setting up, 136
    symmetry, 139
veto, 176-178
point-count tallies, 179-180
pooling equilibrium, 101-102
population paradox, 182
potlatch ceremonies, 118
power
    emotions over rationality, 245-246
    generation example, 168-169

indices
    Banzhaf, 184-185
    defined, 183
    Shapley-Shubik, 185-186
veto players, 176-178
voting proportionality, 182-186
preference patterns over time
    gains and losses, 272
    money, 270-272
presidential election, 186-187
priming, 268
prior probabilities, 30-31
Prisoner's Dilemma
    cigarette advertising example, 70-71
    computer tournament, 312
        Bayesian updating mechanism, 312
        cooperation in competitive environments, 315
        evolutionary pressures, 314
        future payoffs, 314-315
        payoffs, 311
        random defection, 312
        strategies, 312
        tit for tat, 312-313
    cooperation, 67-68, 315
    defecting, 68
    emission reduction, 71
    equilibrium, 68
    experiments, 295
    incomplete information, 85-86
    joint business venture example, 69-70
    overview, 66
    payoffs, 67
    repeating, 305-306
    strategies, 67-68, 312
    strong and weak, 78-79
privatizing versus regulating, 203
probabilities, 19
    expected values, 263-264
    prior, 30-31
    prospect theory, 263-264
    red/blue balls probability game, 257-258
profit-sharing problems, 170-171

property allocation, 155-157
proportionality, 148
   allocation procedures, 157-158
   representation, 132
   utility functions, 257
   voting
      power, 182-186
      problems, 181-182
prospect theory, 259
   behavioral decision-making associations, 259
   elements, 259
   gains versus losses, 259-262, 272
   probabilities, 263-264
   relative values, 262-263
pure strategies
   Nash equilibrium, 55
   zero sum games, 36-38

Q-R

quality
   lemons, 102
   warranties, 105-106

Radiohead pay-what-you-want download, 204
rail line example, 167
Rapoport, Anatol, 200
rationality
   bounded, 298-299
   decision-making, 304
   payoffs, 308
real life versus mathematical theories, 5
reciprocal altruism, 243-244
red/blue balls probability game, 257-258
refrigerator purchase example, 262
regulating versus privatizing, 203
relative values, 262-263
reliability of tests, 31-32

repeating games, 301-302
   deterrence, 304
   equilibrium, 303
   folk theorem, 309-310
   gang of four's cooperation theory, 305-307
   Prisoner's Dilemma, 305-308
   rational decision-making, 304
   supergames, 303
   Ultimatum game, 283
reputations, 112
resource management, 71
restaurant menu pricing, 268
revelation principle, 233
revenue management, 23
   overbooking strategies, 24-26
   strategic choices, 23-24
risk-seekers, 260
rounding electoral votes example, 181
Rousseau, Jean-Jacques, 75

S

saddle points, 48-49
Satterthwaite, Mark, 189
science
   emotional power over rationality, 245-246
   neuroeconomics, 246-248
   neuromarketing, 246-247
   trust, 248-249
Science 84 postcard game, 200
sealed-bid auctions, 207, 213-214
search engine advertising, 229-231
second-price auctions, 209
   generalized, 229-231
   New Zealand airwaves, 212-213
   Vickrey, William, 211-212
selecting
   decision criterions, 22
   strategic choices, 23-24
Selten, Reinhard, 60, 302
sensitivity in tests, 30

separating equilibrium, 100
setting up games
  bargaining, 126, 136
  cooperative, 163
  zero sum, 35
shading bids, 208-209
Shapley, Lloyd, 162
Shapley value
  analysis, 167
  calculating, 166-167
  conditions, 164-166
  power generation example, 168-169
  profit-sharing problems, 170-171
Shapley-Shubik Power Index, 185-186
sharing costs
  mechanism design, 222-226
  power generation example, 168-169
  Shapley value
    analysis, 167
    calculating, 166-167
    conditions, 164-166
  solving, 171-173
  taxi sharing fairness, 161-162
signaling games, 98
  branding, 107
  handicap principle, 107-108
  job market
    analysis, 100
    cost of education, 101-102
    data summary, 100
    overview, 97-99
  nuanced signals
    commitment, 113-114
    defined, 114
    reputations, 112
    spending money, 111-112
    threat strategies, 115-118
    veiled strategies, 118-120
  pooling equilibrium, 101-102
  quality
    lemons, 102
    warranties, 105-106
Simon, Herbert, 298

simple averages, 17
slight market successes, 90
small business versus big business, 302
  deterrence, 304
  equilibrium, 303
  incomplete information, 92-94
  rational decision-making, 304
Smith, Vernon, 277
soccer
  penalty kicks, 292
  rankings, 50
social networking, 316
Social Security digits example, 268
social traps, 201
solutions, 13-14, 194
  cooperative games
    core method, 171-173
    profit-sharing problems, 170-171
    Shapley value, 164-167
  fair division
    adjusted winner procedure, 155-157
    Knaster-Steinhaus Procedure, 151-154
    Last Diminisher Procedure, 150
    proportional allocation procedure, 157-158
    sharing, 155
    three-person method, 148-150
    two-people, 147-148
  Free Rider problem, 204
  incomplete information games, 92-94
  individual temptations
    adverse selection, 220-221
    mechanism designs. *See* mechanism designs
    moral hazards, 221-222
  nonzero games
    credibility, 61-63
    Nash equilibrium, 55-60
  Tragedy of the Commons, 199, 203
  two-player bargaining
    Kalai-Smorodinsky, 143-146
    Nash, 140-143
  Volunteer's Dilemma, 203

zero sum games, 38-39
    CBS example, 45-47
    linear programming, 49-50
    mixed strategy equilibrium, 42
    NBC example, 43-44
    saddle points, 48-49
Sony/Toshiba game, 9
    asymmetric payoffs, 59-60
    mixed strategy, 56-57
    payoffs, 57-59
    pure strategy, 55
spending money signals, 111-112
St. Petersburg Paradox, 251-254
Stag Hunt, 75, 286-287
    analyzing, 76
    payoffs, 75-76
    win-win transformation, 77
stages of trust, 277-278
status quo bias, 273
Steinhaus, Hugo, 151
Steinhaus-Kuhn Lone Divider method, 148-150
stock market example, 273
strategies
    animal kingdom
        evolutionarily stable strategy (ESS), 240-242
        Hawk-Dove game, 238-240
    auctions
        first-price, 207-209
        honest bidding, 229
    backward induction, 279-281
    Chicken game, 73
    choices, 13-14, 23-24
    computer tournament for cooperation, 312
    defined, 4
    dominant, 289-291
    forms
        defined, 6
        Two Finger Morra, 7
    Free Rider problems, 202
    gang of four's cooperation theory, 305-307
    individual versus group, 193
        Free Rider, 201-204
        Volunteer's Dilemma, 200-203
    mixed, 291
        duck pond, 291
        Nash equilibrium, 56-57
        soccer penalty kicks, 292
        zero sum games, 36, 42
    Nash equilibrium, 285
    overbooking, 24-26
    Prisoner's Dilemma, 68
    pure
        Nash equilibrium, 55
        zero sum games, 36-38
    third party candidates, 131
    threat, 115
        brinkmanship, 115-117
        erratic behavior, 117-118
    Tragedy of the Commons
        Hardin's description, 193-194
        high thresholds, 197-198
        low thresholds, 194-197
        moral hazards, 199
        overfishing/deforestation, 194-195
        real-world cooperation, 203
        regulating versus privatizing, 203
        solving, 199, 203
    veiled
        gift giving, 118-119
        losing your temper, 119-120
    Volunteer's Dilemma, 201
    voting manipulation, 189
strategyproof mechanisms, 232
strep throat test, 30-31
subgames
    defined, 63
    perfect equilibrium, 64
supergames, 303
swing states, 185
swing votes, 183
symmetry in two-player bargaining, 139

T

tables
  defined, 6
  Two Finger Morra, 7
tax software example, 262
taxi sharing example, 161-162
technology sharing mechanism, 222-226
tests
  reliability, 31-32
  sensitivity, 30
theater ticket example, 267
theories
  coordination games, 286
  decisions
    defined, 23
    overbooking strategies, 24-26
    strategic choices, 23-24
  folk theorem, 309-310
  gang of four's cooperation, 305-307
  prospect, 259
    behavioral decision-making associations, 259
    elements, 259
    gains versus losses, 259-262, 272
    probabilities, 263-264
    relative values, 262-263
  utility function
    Bernoulli, 253
    von Neumann/Morgenstern, 253-255
  voting, 178
third party candidates, 131
threat strategies, 115
  brinkmanship, 115-117
  erratic behavior, 117-118
Toshiba versus Sony game, 9
  asymmetric payoffs, 59-60
  mixed strategy, 56-57
  payoffs, 57-59
  pure strategy, 55
Tragedy of the Commons
  Hardin's description, 193-194
  high thresholds, 197-198
  low thresholds, 194-197
    analyzing, 196
    payoffs, 197
    three-person, 196
    two players, 195
  moral hazards, 199
  overfishing/deforestation, 194-195
  real-world cooperation, 203
  regulating versus privatizing, 203
  solving, 199, 203
Trivers, Robert, 243-244
true valuations for group benefit
  auctionlike mechanism, 222-226
  generalized second-price auctions, 229-231
  limitations, 231-234
  strategyproof, 232
  Vickrey-Clarke-Groves mechanism, 226-229
trust, 248-249
  Centipede game, 279-280
  conclusions, 29
    Bayes' rule, 29-31
    test reliability, 31-32
  Dictator game, 276-277
  stages, 277-278
Tucker, Albert, 66
TV rights auction, 213-214
Tversky, Amos
  failure of invariance, 267
  mental accounting, 266
  prospect theory, 259
Two Finger Morra, 6-7
two-person bargaining
  anchoring, 268-270
  conditions, 136
  disagreement points, 135
  equality, 137
  feasibility, 138
  individual rationality, 138
  irrelevant alternatives, 139
  Kalai-Smorodinsky solution, 143-146
  linear transformations, 139

Nash solution, 140-143
Nash's standard model, 135
    Pareto optimality, 138
    setting up, 136
    symmetry, 139

U

Ultimatum game, 281-282
    altruism, 282-283
    behavioral study, 281-284
    demographic variables, 284
    neuroscience, 247-248
    physical attractiveness, 284
    raising the stakes, 283
    repeating the game, 283
unconscious intelligence, 300
Upstart versus Goliath, 302
    deterrence, 304
    equilibrium, 303
    incomplete information, 92-94
    rational decision-making, 304
U.S. presidential election, 186-187
used car market game, 102
utilities
    behavior indicators, 255
    Bernoulli's theory, 253
    certainty equivalents, 256
    contradictions, 257-258
    defined, 9
    proportionality, 257
    risk averse, 256
    St. Petersburg Paradox, 251-254
    von Neumann's/Morgenstern's theory, 253-255

V

vacation expenses example, 262
values, 206
    endowment effect, 273-274
    preferences for money over time, 270-272
veiled strategies
    gift giving, 118-119
    losing your temper, 119-120
veto players, 176-178
Vickrey, William, 211-212
Vickrey-Clarke-Groves mechanism
    analyzing, 229
    calculating, 228-229
    overview, 226-228
Volunteer's Dilemma, 200-203
von Neumann, John, 3-4
voting, 131-133
    approval, 187-188
    blocs, 132
    characteristic functions, 176
    as group decision, 175
    impossibility theorem, 188-189
    individual rankings, 179-180
    manipulating, 189
    pairwise comparisons, 178-179
    perfection, 188
    proportionality
        power, 182-186
        problems, 181-182
        representation, 132
    swing states, 185
    swing votes, 183
    theories, 178
    third party candidates, 131
    U.S. electoral system, 186-187
    veto players, 176-178
    weighted point counting system, 131

W-X

war games, 39-40
warranties, 105-106
web
    access example, 130
    advertising, 229-231
    prescription medication purchases, 91
weighted averages, 18-19
weighted point counting systems, 131
win-win outcomes, 77
    Chicken game, 77
    hostages and enforcement, 79
    Prisoner's Dilemma, 78-79
    Stag Hunt game, 77
winner's curse, 210
winner-turns-loser paradox, 180
wireless provider licenses auctions, 214-217
World War II naval battle, 39
Wright, Robert, 54

Y-Z

Zahavi, Amotz, 108
zero sum games
    constant sum, 7-8, 41-42
    defined, 6
    equilibrium solution, 38-39
    misconceptions, 50-51
    setting up, 35
    solutions, 38-39
        CBS example solution, 45-47
        linear programming, 49-50
        mixed strategy equilibrium, 42
        NBC example solution, 43-44
        saddle points, 48-49
    strategies, 36-38
    Two Finger Morra, 6-7
    wars, 39-40

CHECK OUT THESE BEST-SELLERS

More than 450 titles available at booksellers and online retailers everywhere!

978-1-59257-115-4

978-1-59257-900-6

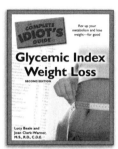
978-1-59257-855-9

978-1-59257-222-9

978-1-59257-957-0

978-1-59257-785-9

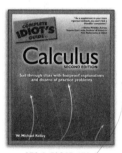
978-1-59257-471-1

978-1-59257-483-4

978-1-59257-883-2

978-1-59257-966-2

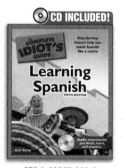
978-1-59257-908-2

978-1-59257-786-6

978-1-59257-954-9

978-1-59257-437-7

978-1-59257-888-7

idiotsguides.com